# Dialysis Patient Care
## Your Questions, Expert Answers

## Lawrence E. Stam, MD

*New York Methodist Hospital*
*The Rogosin Institute*

REVIEWED BY

## Kim Thompson, RN
## Hazel Phillips, RN

JONES & BARTLETT
LEARNING

*World Headquarters*
Jones & Bartlett Learning
5 Wall Street
Burlington, MA 01803
978-443-5000
info@jblearning.com
www.jblearning.com

Jones & Bartlett Learning books and products are available through most bookstores and online booksellers. To contact Jones & Bartlett Learning directly, call 800-832-0034, fax 978-443-8000, or visit our website, www.jblearning.com.

Substantial discounts on bulk quantities of Jones & Bartlett Learning publications are available to corporations, professional associations, and other qualified organizations. For details and specific discount information, contact the special sales department at Jones & Bartlett Learning via the above contact information or send an email to specialsales@jblearning.com.

**Production Credits**

Executive Publisher: William Brottmiller
Executive Acquisitions Editor: Nancy Anastasi Duffy
Medical Writer: Nancy Hoffmann
Editorial Assistant: Jade Freeman
Production Editor: Daniel Stone
Marketing Manager: Jennifer Sharp
Manufacturing and Inventory Control Supervisor: Amy Bacus
Composition: Miranda Design Studio, Inc.
Cover Design: Stephanie Torta
Rights and Photo Research Coordinator: Ashley Dos Santos
Cover Image: Top Right: © Monkey Business Images/ShutterStock, Inc. Top Left: © Photodisc, Center Right: © Daniel Mirer/New York Methodist Hospital, Bottom: © iStockphoto/Thinkstock
Printing and Binding: Edwards Brothers Malloy
Cover Printing: Edwards Brothers Malloy

ISBN 978-1-284-06597-8

2014021067

6048

Printed in the United States of America
19 18 17 16 15    10 9 8 7 6 5 4 3 2 1

# CONTENTS

# The Ten Most Important Things for Dialysis Patients to Avoid

In the thirty-two years I have been working in dialysis, many patients have asked my advice on what to do to remain healthy. We encourage patients to come for their treatments, take medications, and keep their renal diet, but often we neglect to inform them of the pitfalls they must avoid. The concept of predicting what will happen in the future is an important tool of the dialysis nurse. The ability to counsel patients and help them avoid future adverse outcomes can have a major impact on patients' lives.

1. It is important for patients to avoid uncertainty regarding which physician is responsible for their care in the dialysis unit. Often, patients have a primary care physician. The nephrology team that starts the patient on dialysis is often not responsible for the patient in the outpatient dialysis unit. Patients may have dementia or other neurologic conditions that prevent them from remembering their doctor's name. The dialysis nurse should make certain that every patient understands who is responsible for their care in the dialysis unit so that problems can be addressed in a timely fashion. The patient should always be able to turn to that physician for help on medical or other issues.

2. Avoid catheters—patients' number one enemy due to septic and other complications.

3. Avoid short hemodialysis treatments which cause an increase in uremic complications, increased hospitalization rates, and decreased survival on hemodialysis.

4. Avoiding fluid overload is a constant battle to be addressed at every dialysis treatment.

5. Smoking is the number one avoidable cause of cardiac disease and stroke, as well as cancer, in the dialysis patient.

6. Skipping treatments is an all-too-frequent cause of inadequate dialysis and poor survival on dialysis.

7. Obesity is reaching epidemic proportions in both western and developing countries and is a major cause of hypertension and diabetes, as well as destruction of the hips, knees, and ankles that will limit a patient's ability to function independently.

8. Uncontrolled hypertension is important to avoid as part of the prevention of cardiac disease and stroke in dialysis patients.

9. Sedentary lifestyle should be avoided, as it can lead to obesity, decreased muscle mass, and a loss of quality of life in dialysis patients.

10. Noncompliance with medication is a common avoidable cause of poor outcomes in dialysis patients.

*Dialysis Patient Care: Your Questions, Expert Answers* is an easy-to-read discussion of common subjects of importance in the dialysis unit. It is designed to help nurses counsel patients using clear language, which is easy for patients of different educational backgrounds to understand. As the technical sophistication of dialysis treatments improve, the need to form a therapeutic alliance between the dialysis nurse and the patient has never been more important. As physicians spend more time on administrative tasks, the dialysis nurse is playing a central role in medical treatment, education, care and support.

Having dialysis widely available to many patients is one of the most significant achievements of the twentieth century. Improvements in dialysis units in the twenty-first century will result in a better quality of life as well as an increase in patient survival. The forces of globalization and economic development are bringing dialysis to more patients and to areas of the world where dialysis has not been available before. It is a challenging time in the history of this important, life-sustaining therapy.

## AUTHOR

**Lawrence E. Stam, M.D.** is a native New Yorker who graduated from Columbia University and the SUNY/Stony Brook School of Medicine. The material in *Dialysis Patient Care: Your Questions • Expert Answers* comes from his clinical experiences at the New York Methodist Hospital in Brooklyn, New York where he is the Associate Chief of Nephrology. Dr. Stam is an active member of the Rogosin Institute. He is a Clinical Assistant Professor of Medicine at the Weill-Cornel Medical College of Cornell University. His academic interests include teaching students and medical residents. He is a member of the American College of Physicians and the American Society of Nephrology.

# PART ONE

# *Treatment Basics*

What is my role in dialysis treatment?

What medicines are removed by my
patient's dialysis treatment?

*More* . . .

## 1. What is my role in dialysis treatment?

According to the National Institutes of Health (NIH), over 20 million adults in the United States are living with chronic kidney disease (CKD). For patients who progress to end-stage renal disease (ESRD), more than 400,000 receive some form of dialysis treatment, with the majority receiving hemodialysis treatment at a clinic.

Given the many potential complications and importance of successful outcomes, nephrology nurses play a vital role in each patient's dialysis treatment. In fact, each nurse plays multiple roles, which fall broadly into three categories (**Figure 1**):

- Medical treatment
- Education
- Care and support

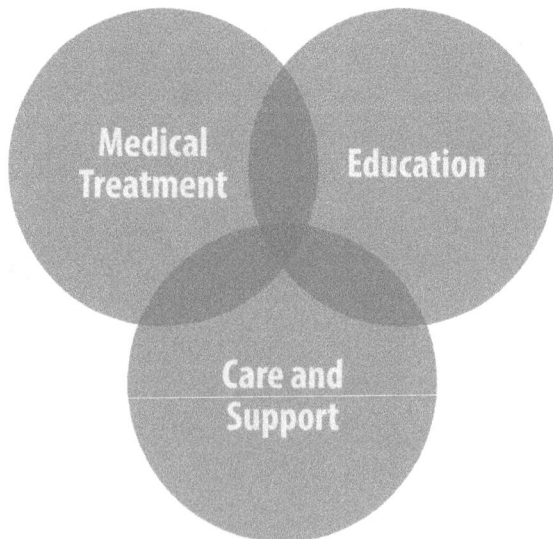

**Figure 1** The nursing role covers three areas of patient care: medical treatment, education, and care and support.

## Medical treatment

As a nurse providing dialysis treatment, you are performing a variety of medical tasks. These include taking patients' vital signs, medication reviews, and monitoring patients' general health. A nurse's role requires great attention to detail, as you must strictly adhere to a set of protocols during the dialysis treatment. Nurses monitor patients during the process and discuss treatment goals and progress as part of the patient's treatment team.

Overlapping the areas of medical treatment and patient education is the topic of infection prevention. The Centers for Disease Control (CDC) has made reduction of healthcare-associated infections (HAIs) one of its primary goals. While the focus was previously on ICUs and other inpatient hospital settings, there is now recognition that this goal is equally important for dialysis centers. In 2008, there were approximately 37,000 central line-associated bloodstream infections (CLABSIs) experienced by hemodialysis patients, according to the CDC. Nurses play an important role in CLABSI reduction, by improving catheter care and washing hands before and after each patient contact.

## Education

Patient education is another important part of your role. You are a resource for your patients and can teach them about renal function, dialysis treatment, diet, comorbid disorders, and general health. Promoting fistula use and talking to patients about good vascular access care will help improve reduction of CLABSI numbers. This book is designed to give you specific suggestions for how to talk to your patients about different aspects of their dialysis care.

### Care and support

Finally, as a nephrology nurse, you must possess great interpersonal skills. Remember, you will be seeing patients on hemodialysis three times a week. Patients will be of all backgrounds, ages, and outlooks. Your task is to educate and motivate all of your patients and to find ways to connect with each one. Because you are seeing them so often, you will get to know your patients and will become a trusted source of support for them all.

Your role as a nephrology nurse is determined by a number of factors. For example, there are professional nursing guidelines to adhere to. There are hospital or center-based protocols to follow. There are state regulations determining what nephrology nurses can or cannot do. There are national initiatives to consider, such as Healthy People 2020 (HP 2020). One of the HP 2020 goals is to reduce the number of new cases of chronic kidney disease as well as associated complications, disability, death, and economic costs. Finally, each patient has individual needs that may also determine the depth of your role in his or her treatment.

## 2. What medicines are removed by my patient's dialysis treatment?

We think of dialysis as a life-saving procedure that removes dangerous poisons and toxins from our patients' blood and bodies. Dialysis also removes beneficial substances, such as vitamins and medications. As long as a substance is water-soluble and its molecules are not too large, there is a good chance that it will pass from the blood through the dialysis membrane into the dialysate. Hemodialysis and peritoneal dialysis membranes are different and have different clearances. For instance, insulin can easily pass through the peritoneal dialysis membrane but cannot pass through the hemodialysis membrane. Many of the water-soluble vitamins, such as vitamin C, folate, and the B vitamins, are removed during hemodialysis and peritoneal dialysis, which is why many dialysis patients are placed on vitamin supplements. If patients eat a healthy, varied diet containing fruits and vegetables, this loss is not significant.

Some antibiotics such as penicillin and ampicillin are removed by dialysis. In general, when you have a patient taking antibiotics, encourage them to take their dose after dialysis treatment. This will decrease the amount of medication removed by treatment. Some anticonvulsant medications are also removed by hemodialysis. In these instances, it is important for patients to have blood levels checked to guide dosing. Other medications removed by dialysis include aspirin and lithium. In fact, overdoses of some medications are sometimes treated with hemodialysis. Alcohol readily passes through the dialysis membrane. There are multiple resources available that detail which medications are removed via dialysis and to what degree. For example, the website www.renalpharmacyconsultants.com has a table with an extensive list of medications and information about whether or not removal during dialysis is likely and if supplemental dosing may be required. Likewise, the American Society of Nephrology has educational information about drug dosing and drug removal during dialysis available on its website, www.asn-online.org.

# PART TWO

# *The Process*

How should my patients prepare for dialysis?

Why does my patient need hemodialysis
three times a week?

What will happen if one of my patients skips a
dialysis treatment or leaves early?

*More . . .*

## 3. How should my patients prepare for dialysis?

In the most optimal situation, patients will meet their doctor before they need to begin dialysis to discuss the process and review blood and other tests. It is important for patients to be given information about dialysis—both verbal and written. The better a patient understands dialysis, the less frightening it will be.

Encourage patients to let both you and their physician know how much information they feel comfortable receiving. Some patients need to hear about every detail. For others, too much information can be overwhelming. Some patients will put themselves in the hands of someone they trust and let them make many of the decisions. Many patients want a family member or friend to come with them and record the information. That person can be their advocate and help them ask questions and make decisions. Some patients learn that they will need dialysis as an emergency and need to begin dialysis right away. This sudden jump into treatment can make the process stressful.

Prior to the start of treatment, the patient's physician may have discussed and created an **arteriovenous (AV) fistula (Figure 2)**. The early creation of the AV fistula can make beginning dialysis easier and safer. Patients should be given plenty of information about why the AV fistula is created, how it functions, and how to care for the site. A **peritoneal dialysis access** or catheter takes less time to be ready than an AV fistula. Most catheters are ready to use in about 2 weeks.

Before beginning dialysis, patients should be given information about their diet (see Questions 8, 9, and 10

**Arteriovenous (AV) fistula**

A method for accessing the bloodstream for dialysis that is created by connecting an artery and vein, usually in the arm, using vascular surgery.

**Peritoneal dialysis access**

A surgical procedure in which a plastic catheter is inserted into the abdomen to allow a patient to undergo peritoneal dialysis.

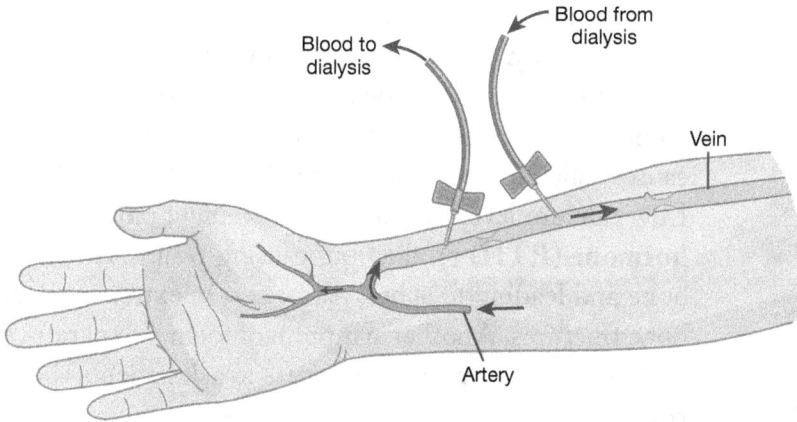

**Figure 2** An AV fistula.

Data from: National Institute of Diabetes and Digestive and Kidney Diseases, National Institutes of Health.

in this book). In many instances, this diet will include a restriction in the amount of sodium and potassium consumed. Phosphorus may be restricted. Phosphate-binding medications can also decrease the phosphorus that is absorbed by the gastrointestinal tract and may be recommended. Many patients mistakenly believe that drinking extra fluid will be helpful in kidney disease. Encourage patients to avoid drinking extra fluid.

Many patients with kidney failure begin to develop anemia as part of their renal failure because the survival of the red blood cells is decreased. Because **erythropoietin** production is decreased in kidney failure, some patients are offered erythropoietin injections (or other medication with similar action) to increase the production of red blood cells and decrease the symptoms of anemia. Counsel patients to be aware of the signs and symptoms of anemia, such as weakness, shortness of breath, and pale skin. Many patients also have to take iron supplements with the erythropoietin injection to increase their red blood cell production.

**Erythropoietin**

A hormone produced by the kidneys that signals the bone marrow to produce more red blood cells. In kidney disease, less erythropoietin is produced and anemia can occur. Erythropoietin can be given by injection to treat anemia.

Due to an inability to eliminate or excrete phosphorus from the body, patients with kidney failure often begin to have bone problems before they need dialysis treatments. This increase in phosphorus in the bloodstream causes calcium to be deposited back into the bones. Low calcium levels in the blood stimulate **parathyroid hormone (PTH)** production, causing re-absorption of bone and leading to aches, pains, and in extreme cases, bone fractures. Another symptom of too much parathyroid hormone is itching, so be sure to ask patients about this symptom.

**Parathyroid hormone (PTH)**

A hormone made and released by the chief cells of the four parathyroid glands found in the neck to regulate calcium metabolism.

Patients with diabetes will continue to have their blood pressure and glucose levels monitored. Many patients with renal failure also have hyperlipidemia and must try to control it through diet, exercise, and sometimes medication.

Leading up to dialysis, patients may feel that it is all a lot of work and not much fun. Many patients find all these problems overwhelming and depressing. Every situation is different. Many patients have no symptoms of anemia or bone disease and make enough urine to not require a very restrictive diet. Remind patients that they are unique, that they must communicate well with you and their physician, and that together you will develop a plan that concentrates on the most important aspects of their care. Patients may feel more confident if they have concrete steps they can follow. Help patients prepare a list of what they may need during their first dialysis treatment. This list should include the patient's insurance cards, current medications, and something to keep themselves occupied (e.g., a tablet, DVD player, or book). Encourage patients to show up 15 to 20 minutes prior to their appointment in case there is any paperwork that needs to be filled out.

## 4. Why does my patient need hemodialysis three times a week?

Hemodialysis removes fluids, salts, toxins, and medications on an intermittent basis only, compared to normal kidney function. Peritoneal dialysis is a continuous therapy and is more similar to normal kidney function. In the early days of dialysis treatment, it was recognized by the medical community that one or two hemodialysis treatments a week were not enough because patients had poor outcomes. Because of these observations, hemodialysis three times a week became the standard therapy. Each dialysis session typically lasts 4 hours, although smaller patients may have shorter sessions and, conversely, larger patients may have longer sessions.

Many patients ask if they can be treated with one or two hemodialysis treatments a week. In the 1980s, physicians tried to decrease the time spent on hemodialysis with treatments called **high-flux dialysis**. This approach resulted in more complications and a shortened life span in patients on dialysis. Experts now recommend that we actually increase the amount of dialysis treatment to prevent complications and improve the quality and length of life.

Recently, there has been an interest in daily hemodialysis. This therapy is more similar to normal kidney function and can be accomplished by patients being on home dialysis with treatments every day. Some patients choose to have home dialysis at night—**nocturnal hemodialysis**, which is performed 3 to 7 times a week. Patients on daily hemodialysis have lower blood pressures, as well as better control of anemia and bone and mineral metabolism because fluid is removed every day. Consequently, they are on less medication, their phosphorus levels are

**THE PROCESS**

**High-flux dialysis**

A procedure that uses high blood flows and large dialysis membranes to remove poisons and toxins in a shorter dialysis treatment, although short dialysis treatments are no longer recommended.

**Nocturnal hemodialysis**

Hemodialysis treatments done at night six or seven days a week at the patient's home while the patient is asleep. Nocturnal hemodialysis treatments utilize a lower blood pump flow and a longer duration of treatment time to achieve gentle treatments with better clearance of toxins and poisons.

lower, and they have more energy. Of course, daily dialysis requires a huge commitment by the patient both in time and training. Many people prefer to have dialysis three times a week because it allows them four days without dialysis treatments. Many patients ask if they will be better off on daily hemodialysis. You can encourage your patients to ask their physician if they can try daily hemodialysis as well as to check with their healthcare insurance to see if nocturnal dialysis is covered. Most patients on daily hemodialysis do not return to three times a week dialysis because they feel better. They are able to eat more varied foods and to increase their fluid intake.

No one therapy is right for everyone. Some patients go from home dialysis to in-center dialysis because of medical problems that require them to be under closer medical and nursing supervision. After their condition improves, they may return to home dialysis.

## 5. What will happen if one of my patients skips a dialysis treatment or leaves early?

Most people on dialysis are confronted with obligations and commitments that interfere with dialysis treatments. A child might be sick. A relative may have a wedding on a dialysis day. A bad cold may make a patient reluctant to leave the house. Things happen at inconvenient times. Bad weather may make travel to the dialysis unit difficult or impossible. Very rarely, power outages may make it impossible for a dialysis center to treat patients for a short period of time.

In most instances, missing or postponing a single hemodialysis treatment does not result in a catastrophe. If patients are having effective dialysis treatments, adhering to their diet and fluid restriction, and taking their medications properly, dialysis can be rescheduled for later the same day or even the next day safely. Patients who are not receiving good dialysis or keeping their diet may develop puffiness around the face, leg edema, and dyspnea due to fluid overload. Typically, these symptoms occur on Sunday or Monday night because of the longer interval between hemodialysis treatments on those days. Often, patients tend to go off their diets on weekends due to parties or other social occasions.

We know that hemodialysis 3 times a week at best replaces 12% to 15% of normal renal function, rather than all renal function. If patients miss one hemodialysis treatment, they miss one-third of their weekly renal replacement therapy. This can result in immediate problems such as excess fluid in the body and hyperkalemia. Weakness, fatigue, and loss of appetite can occur. If dialysis treatments are missed frequently, patients can develop **uremic neuropathy**. Toxins build up at different rates in different patients. For some patients, the build-up of poisons can lead to pericarditis. Other complications include gastritis, congestive heart failure, dysrhythmias, and seizures. **Table 1** outlines potential complications of chronic kidney disease. These complications may arise if patient repeatedly skip dialysis treatments. Signs and symptoms will vary by patient. Patients experiencing the most severe complications may require hospital admission for emergency treatment. It is important for patients to understand how skipping dialysis can impact their health. Part of your role as a patient educator is to explain the importance of sticking to the dialysis schedule.

**Uremic neuropathy**

Damage to nerves that is caused by the toxins and poisons that build up in the blood in kidney failure. Typical symptoms of uremic neuropathy are tingling or numbness in the feet and hands. Uremic neuropathy is treated by increasing the amount of dialysis that the patient is receiving to remove the build-up of toxins and poisons. Medications are sometimes used to treat the symptoms of uremic neuropathy.

**Table 1** Complications of Chronic Kidney Disease

| System | Etiology | Treatment |
|---|---|---|
| General appearance | Tired, weak, sallow skin color due to anemia and toxins | Dialysis and erythropoietin |
| Integumentary | Itching (uremic frost) occurs in an attempt to remove toxins from the body | Dialysis |
| Sensory | Metallic taste in mouth, fishy breath odor (uremic fetor) due to toxins | Dialysis |
| Cardiopulmonary | Hypertension<br>• Related to salt and water retention, erythropoietin (20% of patients on this therapy), or increased renin production<br>• Accelerated renal damage if not controlled<br>• Congestive heart failure develops | • Limiting salt and fluids<br>• Angiotensin-converting enzyme (ACE) inhibitors, angiotensin II receptor blockers, calcium-channel blockers, and beta blockers<br>• Blood pressure goal is 130/80 mm Hg |
| | Pericarditis<br>• Result of metabolic toxins<br>• Chest pain, fever, friction rub, and decreased cardiac output | • Hemodialysis |
| | Congestive heart failure (in 75% of patients needing dialysis)<br>• Result of increased workload of the heart (left ventricular hypertrophy) secondary to anemia, dialysis (shunting of blood), fluid overload, hypertension, and atherosclerosis | • Salt and fluid restriction<br>• Diuretics (loop)<br>• ACE inhibitors and angiotensin II receptor blockers |
| Hematological | Coagulopathy<br>• Platelet dysfunction due to abnormal aggregation and adhesion<br>• Bleeding time increases<br>• May have petechiae or purpura | • Vasopressin replacement (causes release of factor VIII from endothelial cells)—use for active bleeding |

| System | Etiology | Treatment |
|---|---|---|
| Hematological (*continued*) | **Anemia**<br>• Related to decreased erythropoietin production (occurs when glomerular filtration rate falls below 20–25 mL/min) and iron deficiency<br>• Small amounts of red blood cells are lost during hemodialysis | • Erythropoietin if hemoglobin is below 10<br>• Intravenous iron for patients on dialysis (PO absorption of iron is poor) |
| Gastrointestinal | Anorexia, nausea, vomiting, and hiccups—related to metabolic toxins | Dialysis |
| Endocrine | Decreased libido, impotence, and infertility<br><br>**Hypoglycemia**<br>• Damaged kidneys cannot clear insulin from bloodstream | • Dialysis and a healthy diet may restore fertility<br><br>• Patients with diabetes may require lower doses of hypoglycemic agents |
| Mineral metabolism | **Renal osteodystrophy** (disorder of calcium, phosphorus, and bone) leading to bone pain, fractures, muscle weakness, and calcium deposits in the blood vessels, soft tissue, heart, and lungs<br>• As kidney function declines, serum phosphorus increases<br>• High phosphorus causes low calcium levels<br>• Low calcium increases PTH secretion, which causes a high bone turnover | • Restrict dietary phosphorus<br>• Administer phosphorus-binding drugs such as calcium acetate<br>• Vitamin D (suppresses parathyroid hormone) |

*(continues)*

**Table 1** Complications of Chronic Kidney Disease (*continued*)

| System | Etiology | Treatment |
|---|---|---|
| Neurological | Uremic encephalopathy<br>• Appears when glomerular filtration rate falls below 10–15 mL/min<br>• Symptoms: poor concentration (first sign) that progresses to confusion, asterixis, reverse of the sleep/wake cycle, and weakness<br>• Peripheral neuropathy (restless leg syndrome, distal pain, and loss of deep tendon reflexes)<br>• Impotence and autonomic dysfunction | • Dialysis |
| Metabolic | Hyperkalemia<br>• Glomerular filtration rate falling below 10–20 mL/min<br>• Diet high in citrus fruits/juices<br>• Medications such as ACE inhibitors and NSAIDs | • Monitor cardiac status<br>• Dietary potassium restriction |
| Acid–base disorders | Damaged kidneys<br>• Cannot produce enough ammonia or buffer hydrogen ions<br>• Arterial pH generally between 7.33 and 7.37 | • Maintain serum bicarbonate above 21 mEq/L with effective dialysis and sodium bicarbonate supplements |

Patients on peritoneal dialysis are fortunate that they can adjust the timing of their treatment. Then can perform their exchanges at times that are convenient because they are not dependent on the dialysis center schedule. If you are a peritoneal dialysis staff member, you can guide patients on the proper spacing of exchanges. Peritoneal dialysis patients must be self-motivated. If they fail to perform one of their daily exchanges, it is unlikely anyone will remind them. If an in-center hemodialysis patient fails to come for treatment, usually a staff member will call the person to ask why he or she missed a treatment and try to reschedule the treatment. Patients on home hemodialysis can also vary the time of their treatments to fit their schedule. Home dialysis is often a good choice for people with busy careers who may have to adjust their schedules. Still, these patients need to understand the importance of maintaining their daily schedule and the impact that missed treatments will have on their overall health.

Some dialysis centers have monthly attendance awards to reinforce the importance of not skipping treatments. Certificates or small prizes help motivate some patients. Other centers take photos of patients with perfect attendance each month and post them on a wall. Different motivation techniques will work with different patients. Try to remind them that you are all working as a team to achieve the same outcome.

## 6. Can my patient shorten his or her dialysis treatment?

Patients, doctors, and nurses have all always been interested in decreasing the amount of time patients spend on dialysis. Patients, in particular, want to spend as little time on dialysis as possible. It is difficult to sit in one place for 4 hours. The last hour of dialysis is physically and mentally difficult for some patients. A hemodialysis treatment lasting 2½ to 3 hours would be much easier than a 4- or 5-hour dialysis treatment.

High-flux (or high-efficiency) hemodialysis was developed with this goal in mind. The blood flow running through the dialysis machine was increased from 300 mL/min to 450 mL/min. Large dialysis membranes with large surface areas were used to allow more blood to be in contact with more dialysate. Dialysis times were decreased to 3, 2½, and sometimes even 2¼ hours. The owners of dialysis units were encouraged by high-flux dialysis because fewer hours for dialysis meant lower nursing labor costs. However, the results of the experiment with high-flux dialysis were disastrous. Patients had more complications and worse, patients treated with high-flux dialysis did not live as long as patients on regular, longer dialysis. High-flux dialysis was abandoned, and patients went back to longer dialysis treatments.

Several important lessons were learned from this experiment. The importance of increasing blood flow to achieve better removal of poisons and toxins was recognized. The need for better dialyzers with increased clearances and surface areas was appreciated. This finding allowed regular, longer dialysis treatments to remove more poisons and toxins, which made dialysis better by decreasing complications of kidney failure such as neuropathy. In the early days of dialysis, many patients suffered from pericarditis. Some patients developed pleural effusion or **ascites**. These complications are much less frequent now that dialysis is more efficient and removes more toxins.

**Ascites**
A build-up of fluid in the peritoneal cavity.

Recently, nocturnal hemodialysis has allowed patients to receive dialysis treatments while they sleep. This home dialysis treatment involves hemodialysis with a low blood flow rate for 6 to 7 hours, 5–7 nights a week. Patients have much more toxins and poisons removed with this increased dialysis time and feel better. They are able to eat more varieties of food containing potassium and phosphorus that are restricted in the diet of patients on dialysis three times a week. Fluid restriction is not necessary because of daily fluid removal. Blood pressure is decreased by nocturnal dialysis, and some patients can decrease or stop their blood pressure medications. The present trend is to increase the time on dialysis to improve patient outcomes. The best way to decrease time on dialysis is to consider a kidney transplant, which would allow a patient to stop dialysis completely. However, a transplant is not an option for everyone.

## 7. What are some things my patients can do during dialysis treatments?

Generally, your patients' hemodialysis treatments will last 3½ to 4 hours. During part of this time you will be taking the patient's vital signs, reviewing current medications, and discussing treatment. Patients may meet with their physician to discuss treatment and get updated prescriptions. They might talk to a dietitian to learn which foods are healthy to eat and which foods are best avoided. Depending on the dialysis center, patients might also meet with a social worker, podiatrist, or other healthcare professional.

Your patients will probably look forward to chatting with other patients in the waiting area of the dialysis unit and during hemodialysis treatments. They exchange experiences and support each other during difficult times. Many patients on the same dialysis shift have their treatments at the same time for several years, and there is a social aspect to the dialysis unit that is enjoyable for them. Some patients, especially patients who work full time, take the opportunity to nap. Some patients use their laptops to catch up on their work while on dialysis. **Table 2** details some of the ways patients might spend their time during dialysis.

**Table 2** Activities for Patients to do During Dialysis

Work

Nap

Socialize

Watch television

Watch a movie (on laptop or DVD player)

Listen to music

Read (books, magazines, newspaper, catalogs)

Listen to a book on CD or tape

Learn a new language

Catch up on mail/email

Watch/listen to lectures online (college websites, Khan Academy, TED talks)

Crossword puzzles, word puzzles, Sudoku

Call a friend

Knit or crochet

Online games

Complete assignments for online learning classes

Listen to guided meditations

You might also consider group activities. Some centers offer bingo to their patients—either regular bingo or healthcare-based (picture a card with the letters R-E-N-A-L at the top, rather than B-I-N-G-O). Sometimes patients pair up to play chess together—you could even arrange for a center-wide chess tournament. A joke of the day from both patients and staff is something everyone looks forward to. Trivia contests are easy to run for a group, and winners can get prizes or just the glory of being the trivia king or queen for the day!

Patients on peritoneal dialysis spend most of their time performing their exchanges. It takes about 20 minutes to drain the fluid from their abdomen and another 20 minutes to infuse the fresh dialysate. This time can be spent doing other activities.

You might find that some patients start out resenting every minute they are forced to spend on dialysis. You can encourage them to adopt another approach, which is to find activities that are useful and productive, either from a recreational or work point of view. Remind them that dialysis is the perfect excuse for finding some time just for themselves.

THE PROCESS

# PART THREE

# Dialysis and Nutrition

Can my patients on dialysis eat whatever they want?

Why do my patients have to watch what they eat?

How should I explain what a renal diet is?

*More . . .*

## 8. Can my patients on dialysis eat whatever they want?

Before beginning dialysis, patients are often on very restrictive diets to delay the need for dialysis. As renal disease progresses, the kidneys have a harder time eliminating fluid, potassium, and sodium from the body. Often, diuretics are prescribed to increase the urine output and the excretion of fluid, sodium, and potassium. Consuming less sodium, potassium, phosphorus, and fluid compensates for a decreased ability to get rid of these substances. In the early days in the development of dialysis, the limited number of dialysis facilities resulted in many patients delaying dialysis and becoming increasingly debilitated and malnourished. Patients were often placed on low-protein diets and lost part of their muscle mass. They would begin dialysis in a debilitated state and require months of rehabilitation and nutritional support to get back to a state of good health. Now we know that it is better to begin dialysis before these nutritional changes occur and try to preserve patients' muscle mass and nutritional reserves.

An individual's nutritional needs will vary greatly, and it is very important for patients to get expert advice from a registered dietitian who has experience with dialysis patients. This individual evaluation is very important. Every dialysis center has dietary specialists who can take a dietary history and develop a diet that is optimized for each patient.

What your patients should eat will depend on many factors. For instance, comorbid conditions such as diabetes, elevated cholesterol, and lactose intolerance must be considered. Men and women have different dietary needs. Body size and activity level will also determine nutritional needs. The type of dialysis your patients choose will also determine their diet. Finally, the degree of residual renal function each person has will affect their diet.

Hemodialysis three times a week is the most common dialysis therapy. Because this treatment is intermittent, fluid, potassium, sodium, and phosphorus all increase in the body between dialysis treatments. Explain to your patients that during a 4-hour treatment an average of 2 liters of fluid is removed. Fluid will then accumulate in their body until the next treatment. The amount of fluid that can safely be consumed between treatments will include the fluid they excrete in their urine as well as fluid excreted by the lungs and in stool. The goal is to prevent the need for the removal of large amounts of fluid in a short time. If it is necessary to remove a great deal of fluid, complications during dialysis such as muscle cramps, a fall in blood pressure, and dizziness after treatments can occur. Patients will also have to restrict potassium and salt in their diet to prevent them from building up in the body. Protein intake is usually recommended to be 1 to 1.2 grams per kilogram of body weight. Many dietitians have food models to give patients an idea of portion size that is included in their diet. They can also try to incorporate patients' favorite foods into their dietary plan.

For patients on peritoneal dialysis, treatment occurs on a continuous basis every day, enabling the patient to ingest an increased amount of fluid. The treatment can be adjusted to remove more or less fluid by varying the dextrose or sugar content of the peritoneal dialysis solutions. If we wish to remove more fluid, a higher dextrose-containing solution will osmotically pull out more fluid from the body. This will enable patients to drink more in the course of the day. Patients may also eat more potassium on peritoneal dialysis. Protein is excreted in peritoneal dialysis because it passes through the peritoneal dialysis membrane. It is important for patients to take this into consideration and continue to consume enough protein in their diet.

## 9. Why do my patients have to watch what they eat?

The role of diet is probably the most underestimated factor affecting the health of patients. The more we learn of the science of nutrition as patients and health-care professionals, the better prepared we will be to achieve good outcomes on dialysis.

Kidney dialysis patients need expert advice regarding their individual diet. It is likely that many well-meaning people will give them advice, urging them to drink more fluid to flush their kidneys, or to drink cranberry juice. Encourage your patients to thank these well-wishers for their concern, and then discuss diet with their nephrologist, where they will be guided to eat more calories if necessary to gain weight or restrict calories to lose weight. The nephrologist will also advise on fluid intake and diet related to comorbid disorders. Patients will be referred to a renal dietitian to further refine the dietary plan and give specific information regarding the nutritional content of foods.

People with normal renal function are able to eat a large variety of foods without a thought. Damaged kidneys, however, lose their ability to respond to variations in diet and retain rather than eliminate sodium and fluid. Consequently, patients may develop edema, dyspnea, and

DIALYSIS PATIENT CARE: YOUR QUESTIONS, EXPERT ANSWERS

hypertension. Patients with compromised renal function must be reminded to measure and restrict their daily fluid and sodium intake. Dialysis patients must also be concerned with phosphorus. Patients with kidney disease require lower levels of phosphorus intake to protect bone health and prevent **renal osteodystrophy**.

**Renal osteodystrophy**

Demineralization of the bone due to renal failure.

Once patients begin dialysis treatments, fluid, salt, and other substances are removed during the treatment. Patients on hemodialysis have about 2 to 3 liters removed during each treatment (about 2.1 to 3.2 quarts of fluid). Removing more fluid on hemodialysis is not that easy or comfortable for the patient. Therefore, it would be advisable for them to limit fluid intake to the amount that can safely be removed. For patients on hemodialysis every other day, this means decreasing fluid intake to 1 to 1.5 liters a day. Because their kidneys no longer can tell that they are gaining fluid, patients must weigh themselves. For each 1 liter of fluid a patient drinks, their weight increases by 2.2 pounds. Hemodialysis patients routinely weigh themselves before and after their treatments to see if they have had a successful treatment. Peritoneal dialysis patients weigh themselves every day. Because peritoneal dialysis patients are on dialysis every day, they can often drink more fluid than hemodialysis patients on a daily basis. If they drink too much, and their weight increases, they can increase the amount of fluid removed by changing the dialysate. This ability to remove fluid and salt on a daily basis is one of the advantages of peritoneal dialysis.

Patients on dialysis have their potassium, phosphorus, calcium, and cholesterol blood levels monitored and tested monthly. Results should be discussed as a team. Remind patients that by making small changes in their diet in response to these blood tests, they can improve their results. It may seem like a lot of work to some patients, but learning about diet is an important part of dialysis. As time goes on, patients will become more accustomed to their diet and will have an easier time making healthy food choices. Try to teach patients about reading food labels (**Figure 3**) and using available resources (many of which can be found in this book) to help guide their eating.

## 10. How should I explain what a renal diet is?

**Renal diet**

A diet that avoids foods that contain sodium, potassium, and phosphorus, and limits the amount of fluid intake.

A **renal diet** is a term used to describe the guidelines developed for patients with renal disease. Your patient's dietitian and physician will review factors such as age, weight, blood test results, and current urine output in preparing recommendations for an individual diet.

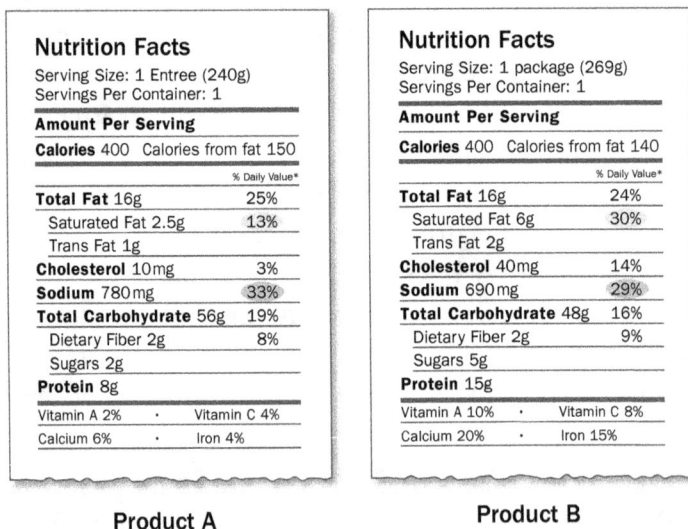

**Nutrition Facts**

Serving Size: 1 Entree (240g)
Servings Per Container: 1

**Amount Per Serving**

**Calories** 400   Calories from fat 150

|  | % Daily Value* |
| --- | --- |
| **Total Fat** 16g | 25% |
| Saturated Fat 2.5g | 13% |
| Trans Fat 1g | |
| **Cholesterol** 10mg | 3% |
| **Sodium** 780mg | 33% |
| **Total Carbohydrate** 56g | 19% |
| Dietary Fiber 2g | 8% |
| Sugars 2g | |
| **Protein** 8g | |

| Vitamin A 2% | • | Vitamin C 4% |
| --- | --- | --- |
| Calcium 6% | • | Iron 4% |

**Product A**

**Nutrition Facts**

Serving Size: 1 package (269g)
Servings Per Container: 1

**Amount Per Serving**

**Calories** 400   Calories from fat 140

|  | % Daily Value* |
| --- | --- |
| **Total Fat** 16g | 24% |
| Saturated Fat 6g | 30% |
| Trans Fat 2g | |
| **Cholesterol** 40mg | 14% |
| **Sodium** 690mg | 29% |
| **Total Carbohydrate** 48g | 16% |
| Dietary Fiber 2g | 9% |
| Sugars 5g | |
| **Protein** 15g | |

| Vitamin A 10% | • | Vitamin C 8% |
| --- | --- | --- |
| Calcium 20% | • | Iron 15% |

**Product B**

**Figure 3** Comparing product labels.

Most likely, a renal diet will restrict salt or chloride intake to 2 grams (2,000 milligrams) a day to reduce the risk of hypertension. Hypertension can put a strain on the heart or lead to a stroke. If the heart becomes over-loaded by too much salt and water, it may fail to pump enough blood, causing shortness of breath as fluid backs up into the lungs. Potassium must also be restricted to 2 grams, or 2,000 milligrams, a day. High potassium levels may cause heart problems and muscle weakness. Remind patients to be cautious about eating oranges, orange juice, cantaloupe, bananas, and tomatoes, as they are all high in potassium.

Patients also need to understand the importance of phos-phorus. Phosphorus is an important mineral because it is needed to store energy from food for later use. It is also present in our bones. The only way we can excrete phos-phorus from our body is through our kidneys. As our kidney function decreases, phosphorus builds up in our bloodstream, causing calcium in our blood to be depos-ited back into our bones. As the blood-calcium level falls, it stimulates the production of parathyroid hormone by the parathyroid gland. Parathyroid hormone will cause more calcium to be absorbed in our small intestine. It will also cause the release of too much calcium from our bones, which can lead to weakening of our bones and even fractures. We must control the phosphorus level in our blood to keep our bones healthy. This is accom-plished by taking medications called phosphate bind-ers that bind phosphorus that we eat in our intestine. We also must decrease the amount of phosphorus we eat every day. Educate your patients about phosphorus-containing foods, like meat, dairy products, and beans, such as lentils and kidney beans.

Protein is also an important part of the renal diet. Patients need to eat about 1.2 grams of protein per kilogram of body weight, which is 7 to 8 ounces of meat, chicken, or fish a day. Many people are under the impression that they should not eat protein if they are a kidney dialysis patient. This advice is wrong and could lead to protein malnutrition. Explain to patients that protein is needed to build and repair muscles, make blood cells, fight infection, and perform many other metabolic tasks in our body.

Patients may want to know how many calories are needed every day. Obviously, this depends on age, physical activity level, and weight. About 30 kilocalories per kilogram, or 2,000 to 2,500 calories, will be the ideal daily range for most kidney dialysis patients.

Individual diets will also depend on comorbid medical conditions, such as diabetes or hyperlipidemia. When patients begin dialysis and are learning about diet, they may be given a diet sheet. Some patients may see this sheet and focus on their favorite foods that are prohibited, increasing their feeling of loss. Not only will they be stuck on dialysis, but they will never be able to eat normal foods again. I prefer to think of the renal diet as a voyage. We start in one place and slowly travel to another place. Many patients have some residual renal function when they start dialysis. This means that they are making urine and getting rid of some toxins and poisons, but not enough to stay off dialysis. The renal diet is not as important at the beginning of dialysis as it will be later on when the residual renal function is gone. Most patients have time to adjust to the renal diet. At the beginning of dialysis treatments, patients should concentrate on one or two important points. This may be to decrease fluid and salt intake, or if they already eat a low-salt diet, then choose two other important points. This makes the process more manageable and will prevent patients from getting overwhelmed. They will head in the right direction and develop healthier eating habits.

## 11. Can my patients on peritoneal dialysis eat more foods?

Yes, one of the joys of peritoneal dialysis is the ability to eat more varieties of food. For instance, patients on hemodialysis must restrict their diet to about 1 to 1½ quarts or liters of fluid a day. This is because hemodialysis is an intermittent therapy. This means that patients receive a treatment and then fluid and poisons build up in their body for 2 days before the next treatment. However, in peritoneal dialysis, patients receive treatment throughout the day, every day. Therefore, fluid is removed every day. Potassium is usually a little low or normal, allowing patients to consume orange juice, potatoes, tomatoes, and other high-potassium foods. Phosphorus, which is restricted on hemodialysis, is also removed on a daily basis on peritoneal dialysis. Patients on peritoneal dialysis can eat more dairy products such as milk, ice cream, and cheese, as well as beans and meats.

One would think that everyone would want to be on peritoneal dialysis so they could avoid a restrictive renal diet. However, peritoneal dialysis is less efficient than hemodialysis. This means that very large patients are not able to remove enough toxins and poisons to prevent the complications of renal failure. Patients who do well on peritoneal dialysis have some residual renal function. Their kidneys still produce some urine and this helps eliminate more poisons. Another consideration is whether a patient is willing to be on dialysis every day. Peritoneal dialysis patients do their own dialysis. This allows flexibility but also requires responsibility. Your patients may ask you about the pros and cons of peritoneal dialysis. It may also be helpful for them to meet patients on peritoneal dialysis so that they can hear in these patients' own words the advantages and disadvantages of being on peritoneal dialysis.

It is important to mention that daily nocturnal home dialysis also allows the patient to eat and drink a larger variety of foods. This is because dialysis is typically done 6 or 7 days a week. Fluid, salt, potassium, phosphorus, toxins, and poisons are all removed on a daily basis. This allows patients to eat at their favorite restaurant, then go home and remove all of the substances that can cause problems just like fully functional kidneys would. Some patients may want to consider daily nocturnal dialysis.

Patients with renal failure who choose to have a kidney transplant have an increased variety of food choices. Many transplant patients must still restrict the amount of salt they eat because they may still tend to retain salt. They can drink more fluid and eat more potassium and phosphorus than hemodialysis patients. They must still pay attention to their cholesterol intake if they have a high cholesterol level. Diabetic patients will still be diabetic when they are on renal replacement therapy. Many diabetic patients will have a better appetite when they begin dialysis, resulting in an increased intake of calories and an increased blood sugar. It is important to remind patients that they will need to adjust their diabetes medications after starting renal replacement therapy.

## 12. How much fluid should my patients drink every day?

The amount of fluid that patients can safely drink on dialysis will depend on the amount of urine they make. It is a lucky patient who continues to make a large amount of urine, because it allows them to drink more. Before a patient begins dialysis, their physician will probably order a 24-hour urine collection for creatinine clearance. This test will help determine if the patient needs to start dialysis and will give an idea of how much urine they are producing every day. Most people have no idea. We know how many times we urinate but not how much. You can explain to patients: If you are making a quart of urine a day, you can add that amount to the 1 to 1½ quarts of fluid you should drink each day.

On a daily basis, it is too much work to measure every ounce of fluid a person drinks. You can suggest that patients try this useful exercise once or twice, but life is too short to measure every drop that goes into our bodies! Fortunately, if someone gains 1 kilogram of weight, that corresponds to 1 liter of fluid consumed. (Some dialysis units use pounds of weight, but most use kilograms when they weigh patients before and after dialysis.) This formula enables us to determine if someone is drinking too much. If a patient drinks 1 liter of fluid and urinates 1 liter of fluid, they will lose weight. This seeming imbalance is due to **insensible fluid loss**. Insensible fluid loss is fluid lost in bowel movements in the form of stool, in sweat, and through water vapor from the lungs. Insensible loss is about one-half to three-quarters of a liter of fluid a day. So, we measure

**Insensible fluid loss**

The fluid that leaves the body in the form of water vapor from the lungs, sweat, and fluid in stool. The amount of insensible fluid loss is about 2 to 3 cups every day.

urine output, add one-half a liter a day of insensible loss of fluid, and another liter a day of fluid that is to be removed on dialysis, and we get a figure that is roughly 1½ to 2 liters of fluid a day. You can tell patients: Just weigh yourself after dialysis and expect to gain 2 kilograms (or 4.4 pounds) before your next treatment. Either method will work.

Many patients on dialysis are very thirsty. The exact reason for this is not known. It may be due to the effect of hormones released by the body, or it may be a signal from the brain that causes patients to seek water to drink. Some medications for blood pressure such as clonidine can cause dry mouth. Many patients resort to eating ice to decrease their thirst. Unfortunately, melted ice is water, and this is not an ideal solution. It is important for patients to have some sugar-free gum or candy to pop in their mouth when they become thirsty. This will decrease the amount of fluid they consume. It is also important to teach patients about the fluid content of foods. The best example would be watermelon, which contains a large amount of fluid, but other examples include grapes, celery, lettuce, and oranges.

## 13. Is it okay for my patients to eat protein while on dialysis?

Previously, many people believed that eating protein was harmful to the kidneys and therefore unhealthy for patients on dialysis. In the early days of treatment for kidney failure, before dialysis, physicians discovered that reducing the protein intake of patients allowed them to live longer because they produced fewer poisons and toxins on a low-protein diet. Patients became malnourished but had fewer symptoms of kidney failure. In the 1980s, a series of experiments with rats showed that rats with kidney failure lived longer on a low-protein rat chow diet than rats with kidney failure on a regular-protein rat chow diet. This research rekindled the interest in low-protein diets in kidney failure. Studies in humans failed to show a clear benefit of low-protein diet in patients with kidney failure. Now, the only role for reducing protein in the diet may be in the pre-dialysis period. Some physicians feel that placing patients on a 30- to 40-gram protein diet may help keep them off dialysis for a time. Not all physicians recommend this, though.

The danger of decreasing protein in the diet is the development of protein malnutrition, leading to **muscle wasting**. This condition can result in weakness, fatigue, anemia, and a decreased ability to repair our body when injured. It can also lead to a decreased ability to fight infection. If any of your patients is considering a low-protein diet for kidney failure, they should be observed carefully for malnutrition. This is especially important in children, elderly patients, and patients who are losing large amounts of protein in the urine.

Most doctors and nutritionists recommend that patients with kidney failure eat a good amount of **high-biologic-value protein** each day. It is unlikely that your patients have heard this term, so explain to them that high-biologic-value protein is protein that contains all of the essential amino acids that we need and that can be easily absorbed by our digestive system. Vegetarian patients may argue that many vegetables contain protein; however, they do not contain all of the essential amino acids needed for good nutrition. **Table 3** lists examples of healthy protein choices. Patients who are vegetarians must take special care to eat a variety of foods that will provide

**Muscle wasting**

A decrease in the size of the muscles of the body due to poor nutrition or medical illness.

**High-biologic-value protein**

Proteins found in foods such as eggs, fish, poultry, soy products, and meat. High-biologic-value proteins are easily digestible and are rich in the ten essential amino acids that the body cannot make.

them with all of their protein requirements. This means eating 1.2 grams of protein per kilogram of body weight. For non-vegetarians, this nutritional requirement can be satisfied by eating 4 ounces of meat, poultry, or fish twice a day. Most of us have no idea of the weight in grams or ounces of the food we eat. If your center has a renal dietitian, they can help patients determine the portion size of food that is needed for good nutrition. Food models can help patients understand portion size, or you can use the guide in **Table 4**.

If your patient cannot seem to eat enough protein, you can suggest protein supplements in a powder or liquid form. Many patients have a poor appetite in the pre-dialysis period because poisons and toxins are building up in their body, and they lose interest in their favorite foods. It is interesting that some patients selectively decrease the amount of protein they eat. These patients may be trying to decrease the amount of poisons and toxins they produce.

Patients in the pre-dialysis period often lose weight and some of their protein stores. They may suffer from nausea, vomiting, and hiccups, which interfere with

**Table 3**  Healthy Protein Choices

| |
|---|
| Lean meat — baked or broiled chicken, turkey |
| Fish — water-packed tuna, trout, cod |
| Cottage cheese with berries |
| Sliced hard-boiled egg whites |
| Protein bars |
| Yogurt with granola |

**Table 4** Playing with MyPlate Portions

Your favorite sports and games can help you visualize MyPlate portion sizes.

| GRAINS | 1 cup dry cereal | 2 ounce bagel | ½ cup cooked cereal, rice, or pasta |
|---|---|---|---|
| | | | |
| | 4 golf balls | 1 hockey puck | tennis ball |

| VEGETABLES | 1 cup of vegetables |
|---|---|
| | |
| | 1 baseball or 1 Rubik's cube |

| FRUITS | 1 medium fruit (equivalent of 1 cup of fruit) |
|---|---|
| | |
| | 1 baseball |

| OILS | 1 teaspoon vegetable oil | 1 tablespoon salad dressing |
|---|---|---|
| | | |
| | 1 die ( $^{11}/_{16}$" size) | 1 jacks ball |

*(continues)*

**Table 4** Playing with MyPlate Portions (*continued*)

| | | |
|---|---|---|
| **MILK** | 1 ½ ounces of hard cheese | ⅓ cup of shredded cheese |
| | 6 dice (¹¹⁄₁₆" size) | 1 billiard ball or racquetball |
| **MEAT AND BEANS** | 3 ounces cooked meat | 2 tablespoons hummus |
| | 1 deck of playing cards | 1 ping pong ball |

eating. These symptoms usually resolve within several weeks of dialysis treatments. Infections occurring in patients on dialysis can lead to protein malnutrition. Examples of this would be foot infections, access infections, or pneumonia. Attention to adequate nutrition during these episodes of infection can help speed their recovery and return them to their normal routine.

## 14. Can my patients eat during their dialysis treatments?

Peritoneal dialysis patients can eat during treatments because dialysis is taking place 24 hours a day. The dialysis usually does not interfere with the ability to eat an adequate diet. Peritoneal dialysate contains **dextrose** to help remove fluid from the bloodstream into the peritoneal space so it can be discarded with the dialysate. Some dextrose is absorbed by the body. These extra calories from the dextrose can sometimes act as an appetite suppressor and decrease the patient's desire to eat. Occasionally, patients feel full when their abdomen is full of dialysate. This symptom can be managed by having the patient eat when the abdomen is drained of dialysate.

Hemodialysis centers have different policies regarding patients eating on hemodialysis. Some centers are concerned about the risks of patients choking or potential hygiene issues. Obviously, you must follow your center's policies. A good guideline is to recommend that patients not eat during hemodialysis in the first few weeks of treatment. Many patients have episodes of vomiting as the high levels of poisons and toxins are removed during their first hemodialysis treatment, so they definitely will not want to have a full stomach for that. After adjusting to hemodialysis, many patients find that eating a small snack during their treatment is enjoyable. This may be especially important in patients who are diabetic and who need to eat to regulate their glucose. One of the most important points to make to patients is for them to choose low-phosphorus foods. **Table 5** lists a few foods that are lower in phosphorus. Encourage your patients to get creative in the way they make snacks. For instance, rolling up some lean chicken in a flour tortilla

**Dextrose**

A sugar added to peritoneal dialysis solutions to help remove fluid from the body. It is also added to intravenous fluids to provide calories for nutrition.

**Table 5** Examples of Low-Phosphorus Foods

Green peas

Refined grains (crackers, cereals, bagels)

Lean pork or poultry

Frozen fruit pops

Popcorn

Canned tuna

Flour tortillas

Couscous

Plain white rice

Shrimp

Applesauce

Peaches

Unsalted pretzels

with some feta cheese and a sprinkling of mixed greens would make a healthful, low-phosphorus snack.

Patients on home hemodialysis frequently combine dialysis with their dinner plans. They come home from work, receive their dialysis treatment, and enjoy having dinner with their family while they are receiving their treatment. This can help with a busy schedule and create time for other activities.

## 15. Why are potassium and phosphorus so important, and how can I best explain their impact on patients?

Potassium and phosphorus occur inside and outside the cells in our body and both needed for healthy living.

Potassium and phosphorus levels in the blood reflect the quality of the dialysis treatment and how successful patients are in following the renal diet.

Most of the potassium is inside our cells, though it is also present in our blood. We eat potassium-containing foods daily and excrete the potassium through urine, stool, and sweat. In patients with kidney failure, the ability to eliminate potassium in the urine decreases. In order to compensate, they must decrease their potassium intake through food and increase their elimination of potassium with the help of the dialysis treatment. We cannot measure the potassium inside the cells of our body, so we use serum potassium levels to determine if a patient's potassium level is optimum. A potassium level that is too high or too low can cause panic in the dialysis unit. This is because very high and very low levels can be associated with heart problems, such as irregular or slow heartbeats and even cardiac arrest. Muscle weakness and even paralysis are other signs of an abnormally high potassium level. Most patients with these levels do not develop such problems, but it is important to correct a low or high potassium level as soon as possible.

The most common cause of a high potassium level is eating too much potassium in the diet. Orange juice, bananas, cantaloupe, dairy products, tomatoes, salt substitutes, and sports drinks such as Gatorade have too much potassium in them to be safe foods for the hemodialysis patient. The next common cause of a high potassium level is skipping dialysis treatments. An elevated potassium can also be due to a problem with the dialysis treatment. If the access that supplies blood to the hemodialysis machine is not working properly, not enough blood will pass over the dialysis membrane, and a decreased removal of potassium occurs. If there is a blockage in the access, the same blood can go from the patient to the machine over and over again without mixing with the rest of the blood in the body. This is called **recirculation** and is prevented by a frequent evaluation of the AV access. Patients can also liberate too much potassium from the organs in their body due to disease. This can occur during crush injuries when damaged muscle lets large amounts of potassium that is usually found inside the cells to leak out into the bloodstream. Another cause of a high potassium level is bleeding in the gastrointestinal tract from an ulcer or gastritis. Red blood cells contain large amounts of potassium, which can be absorbed from the bowel.

The immediate treatment of a high potassium level in a dialysis patient should be to place the patient on dialysis as soon as possible. At the same time, an evaluation should be made of the cause of the high potassium and steps should be taken to decrease the potassium intake in the diet. Low potassium can be the result of vomiting and diarrhea, laxative use, or a decrease in appetite leading to a very low intake of potassium over a long period of time. This problem can often be managed by changing the potassium level in the dialysate and improving the patient's nutritional intake.

Phosphorus, like potassium, is in the food we eat and is eliminated from the body in our urine. As our kidney function decreases, the level of phosphorus in the blood begins to rise. This can be managed by decreasing the amount of phosphorus eaten. High-phosphorus foods include meat, dairy products, beans, and cola drinks. In addition to eating a reduced phosphorus diet, most dialysis patients will require phosphate binders. These medicines are taken with food and will bind the phosphorus in the gastrointestinal tract and prevent the phosphorus

**Recirculation**

Occurs during hemodialysis treatments when blood that has gone through the dialysis machine does not mix with the patient's blood but returns to the dialysis machine.

from being absorbed. If the phosphorus level is too high, patients can complain of severe itching. In addition, the high phosphorus will cause the calcium in the blood to be deposited in the bones, causing the calcium level to fall, which will stimulate the production of parathyroid hormone (PTH). The parathyroid hormone, if present in high levels, can cause bone disease accompanied by pain or even fractures. Although a high phosphorus level is not as dangerous as a high potassium level, long-standing high phosphorus can affect the quality of a patient's life on dialysis and should be avoided. Low phosphorus is uncommon in dialysis patients. If your patient's phosphorus is low, consult with the nephrologist about having the patient stop taking his or her phosphorus binders. A small increase in dietary phosphorus may also be needed.

## 16. How can I help my patients improve their appetite?

Kidney patients on dialysis have many reasons for having a poor appetite. Poisons and toxins that build up in the body of a patient with kidney failure change the way food tastes and smells. Many patients report a metallic taste in their mouth from the build-up of poisons and toxins in the body. Kidney patients may have a dry mouth, which may cause them to drink large amounts of fluid with no nutritional value, leaving them too full to eat solid foods with more nutrition. Nausea, vomiting, and hiccups are common symptoms of kidney failure in the pre-dialysis period. Patients may become afraid that eating will cause them to vomit.

Beginning dialysis will reverse some, but not all of these symptoms. Many patients who delay beginning dialysis will spend several months with poor nutritional intake. As they lose weight, they become physically weaker and less mobile. They suffer from nausea, a decreased appetite, and general malaise. These symptoms decrease their interest in and ability to consume an adequate amount of food and so they become accustomed to eating less. When they begin dialysis, they have to get in the habit of eating more again. Patients are often overwhelmed by the renal diet that is presented to them and may be unsure of what is safe for them to eat. All patients who begin dialysis have an element of depression present. Patients often feel that beginning dialysis is a loss and that life as they knew it is over. This depression is often associated with a decreased appetite.

Fortunately, helping patients to improve appetite and nutrition is not as hard as it seems. The first thing for patients to realize is that diet is equally as important as dialysis treatments and medication in getting well. Eating is a social experience. We celebrate special occasions like birthdays and holidays with a special meal. People generally will eat more if they share their meals with family and friends. If they are in the hospital, suggest that patients have a friend or family member visit at mealtimes. They can bring a sandwich so that they eat with the patient, which will help distract the patient from viewing eating as a task to get through, and ideally reconnect them with food as a pleasurable experience. Encourage patients to eat five small meals or snacks a day. Patients who fill up on small amounts of food can let their stomachs empty a bit before they eat again. Remind patients that if they need to consume more calories, they should not eat or drink anything with the words "diet" or "low-fat" on the label. What is a good idea for well-nourished individuals may not be the best food for malnourished people. Teach your patients to make sure that everything they consume you has some caloric value. Recommend drinking a liquid supplement or apple juice, not water, if they are thirsty and remind them to avoid carbonated beverages. The bubbles can distend your stomach and make you feel full, leaving less room for food.

Physical activity is one of the best ways to increase your appetite. Active people consume a great deal of calories. They are less likely to suffer from constipation and other bowel complaints. Give your patients the following tips:

- Try to take a short walk several times a day.
- Find a reason to go out even if you do not have to.
- Buy a newspaper or pick up a quart of milk or juice.
- Resist the urge to sit and watch television.
- Do some work around the house or the garden, even if it is only a small task that can easily be done by someone else.

All these activities will help patients to increase their appetite.

Medications are sometimes used to increase appetite, as well as over-the-counter zinc supplements. Patients with zinc deficiency can lose their taste for foods, so the supplements help regain appetite. Antidepressants (specifically, serotonin reuptake inhibitors) can help some depressed patients with kidney failure gain weight. Although medications are usually not an easy answer to the problem of poor appetite, they may be useful in selected patients.

## 17. How can my dialysis patients lose weight?

There are a few different types of weight to discuss with dialysis patients. The first, which is the weight lost on dialysis, corresponds to the fluid removal. Another is called **dry weight**, which corresponds to weight after dialysis when there is no extra fluid in the body. Dry weight is the target weight to aim for where patients feel well and have good blood pressure control. The third weight is called **ideal body weight**. This weight is dependent on age, height, and body build. Ideal body weights are different for men and women and are based on population studies. Use current tables to help patients find their ideal body weight.

Being overweight may have several medical consequences, including elevated blood pressure, difficulty controlling diabetes, cardiovascular stress, joint stress (knees and hips from carrying excess weight), and sleep apnea. For all these reasons, kidney dialysis patients who are overweight are often encouraged to lose weight.

Losing weight as a dialysis patient is similar to losing weight in people not on dialysis: It requires commitment. Decreasing daily caloric intake and increasing activity level with exercise is the standard approach. Dietitians at dialysis units can help patients identify foods that have high nutritional value and fewer calories. Many different diets are advertised in the media for losing weight. Many of these diets are very restrictive. People frequently lose a lot of weight in a short time only to gain it back. Diets that aim for a slower loss of weight by changing eating habits are more likely to be effective long term. Remind patients that it is important to check with their nephrologist and dietitian before

**Dry weight**

The body weight of hemodialysis patients at the end of their hemodialysis treatment. The dry weight is determined by removing fluid until the blood pressure begins to fall.

**Ideal body weight**

Ideal body weight is dependent upon age, height, body, build, and sex, and is based on the measured average weight of normal individuals. Ideal body weights can be found in published tables, which are available from a kidney dietitian.

starting any diet to see if the diet is safe for kidney dialysis patients. Diet pills are generally not a good idea for dialysis patients, as they can have some serious side effects. In extreme cases of obesity, weight-loss surgery called bariatric surgery can be performed to band the stomach, causing a decrease in the ability to eat too much. Intestinal bypass operations may also be used to treat obesity. These procedures are used as a last resort when dieting and medical management has failed.

Most people find it difficult to lose weight alone. Share these tips with your patients:

- Support groups such as Weight Watchers have been successful in helping people lose weight because they have a structure and a great deal of positive support. It is easier to keep your diet when you have friends who are also trying to lose weight.

- If your spouse, significant other, or other family members are also overweight, consider encouraging them to join you in your weight loss program. This may make it easier to change your eating habits and to limit some of the high-calorie food that comes into your home.

- If you do buy high-calorie foods as a treat, consider buying a smaller container.

- Avoid extra-large sizes of cookies, chips, candy, and cakes, even if they cost the same as the small package. This way if you go off your diet, you will not eat as much of these foods.

- Exercising is also easier to do with others. If you have a regular appointment with a friend to walk, jog, bike ride, or go to the gym, you are more likely to exercise.

The best results in losing weight are because of changes in a person's lifestyle that result in a gradual loss of weight.

## 18. Who should provide my patients with nutrition guidelines?

The renal dietitian associated with your dialysis unit will be able to help patients with their nutrition. Patients on dialysis come from many different cultures and backgrounds. They are used to eating ethnic foods that may or may not be appropriate for patients with kidney failure. Often the renal dietitian can recommend substitutions in the recipe of a favorite food to make the food more compatible with the renal diet.

The tools that the renal dietitian will use to evaluate a patient's nutritional state are the monthly blood tests, most importantly the serum **albumin**, which reflects the protein stores in a patient's body. The albumin level can be low for a variety of reasons. A patient:

**Albumin**

A water-soluble protein found in the blood normally in very minute amounts. Patients with kidney disease often have albumin in their urine because the albumin leaks through the damaged kidneys into the urine.

- may be malnourished and not eating enough protein
- may be losing albumin through their urine due to kidney disease
- may be suffering from liver disease and may not be able to build enough albumin protein despite adequate protein intake.

The other important lab values include BUN and creatinine, phosphorus and calcium levels, and cholesterol level. If a patient's cholesterol level is too low, it may be a sign that the patient is not eating enough food. The renal dietitian can use these tests along with height and weight to give a nutritional snapshot of where each patient is at.

Remind patients that the advice of the renal dietitian is not an attack on food choices. It is rather an observation on how food choices will affect overall health in the long run. The renal diet is a voyage, with patients starting in one place and hoping to arrive in another, healthier place. As a nurse, you and the renal dietitian are fellow travelers with your patients.

The American Association of Kidney Patients, www.aakp.org, has a wonderful publication called *Beginnings, Living Well With Kidney Disease*. This publication features many excellent articles concerning nutrition and recipes for great meals, and it is an enjoyable resource for patients on kidney dialysis. Other great recipes can be found at www.culinarykidneycooks.com and www.kidney-cookbook.com.

# PART FOUR

# Lifestyle Changes

Since being on dialysis, my patient feels tired all of the time. Will this improve?

How do I help my patients manage their fatigue?

Can my patients smoke on dialysis?

*More . . .*

## 19. Since being on dialysis, my patient feels tired all of the time. Will this improve?

It seems that patients who have been on dialysis long-term, whether hemodialysis or peritoneal, tend to feel the most fatigued. The reasons for this are not definitively clear. Some healthcare providers suggest that the fatigue is due to anemia, malnutrition, and the overall effect of dialysis on the patient. However, some research refutes these assumptions. Certainly the stress of having a chronic illness contributes to a patient's feelings of fatigue.

Patient fatigue may also be due to the decline in kidney function they have experienced. One study found that patients with higher serum creatinine levels at the beginning of dialysis had higher vitality (energy) levels than their cohorts one year later.

The factors leading to patient fatigue might be out of your control and even the patient's control. However, managing the fatigue is definitely something you and the patient can tackle together (see Question 20).

## 20. How do I help my patients manage their fatigue?

Energy levels vary among individuals both before and after starting dialysis. We all know people who are in constant motion. They work hard, find time for a variety of activities and hobbies, and are always ready to go off on another endeavor at a moment's notice. The rest of us feel tired just thinking about doing these things. It is important to realize that dialysis is physically and psychologically demanding. It is important for patients to take extra care of themselves during the beginning stages of dialysis treatments. Suggestions to pass along to patient include: getting a full night's sleep; sitting down and eating three meals a day and a snack even if they aren't hungry; avoiding extra life stressors like moving, taking on a new job, or divorcing a spouse. These experiences can take a lot out of any of us and add to overall fatigue levels.

Patients who feel tired should review medications with their physician. Many medications for blood pressure and pain, and some ulcer medications called histamine $H_2$-receptor antagonists, have side effects that may make patients feel tired. Patients on dialysis are more likely to have side effects from medications, possibly because they metabolize medications more slowly than patients with normal renal function. Sleeping pills can build up in the body over time and cause drowsiness. Medications prescribed for anxiety and depression can sometimes decrease energy levels.

Because kidney failure causes the sleep-wake cycle to be reversed, many patients with kidney failure have difficulty sleeping at night. Patients may instead fall asleep during the day. This problem is due to changes in the body's hormonal cycle. Better sleep at night can be achieved by avoiding naps during the day. Let patients know that it is a good idea not to drink caffeinated beverages in the hours preceding sleep.

Finally, it has been shown that low to moderate levels of exercise improve energy levels and relieve daytime fatigue. Patients who are feeling fatigued may not be inclined to start an exercise program because they may feel too tired to do so! It is important to encourage them to start off slowly and remind them that they will soon see the benefits of exercise. Walking around the neighborhood or a nearby park 2-3 times a week for 15 minutes could be a starting point. From there, patients can increase their activity level depending on their individual circumstances. Assure patients that a moderate exercise program will have far-reaching benefits for their health.

## 21. Can my patients smoke on dialysis?

A significant number of patients who smoke before they begin dialysis continue to smoke after they begin dialysis. All hemodialysis centers prohibit smoking while on dialysis. Many patients are surprised to learn that smoking cigarettes is an independent risk factor for kidney disease. This risk is because cigarette smoking is one of the causes of **atherosclerosis**. Involvement of the arteries to the brain can lead to stroke. Atherosclerosis occurring in the arteries to the kidneys results in a decreased blood supply to the kidney. When atherosclerosis occurs in the coronary arteries, it leads to coronary heart disease. Atherosclerosis is also caused by hypertension, diabetes, and high cholesterol, and occurs more frequently in patients who have other family members with the disease. Because kidney patients have a higher chance of having hypertension and high cholesterol, it is especially important that they try to stop smoking. Diabetic patients on hemodialysis are especially vulnerable to the effects of cigarette smoking. There is a significant decrease in 5-year survival in hemodialysis patients with diabetes who smoke.

Good educational points to share with patients include:
- Most smokers want to quit smoking, and every year about half of them try to quit.

**Atherosclerosis**

A disease characterized by thickening and hardening of the arterial wall due to plaque build-up.

- Within 2 days of stopping, the ability to smell and taste food improves.
- Within 2 to 3 weeks, the chance of a heart attack decreases, and circulation and breathing begin to improve.
- The Surgeon General has declared smoking to be the number one preventable cause of death and illness in the United States.

Try the following step-by-step approach with patients:

1. The first step in quitting smoking is to set a quit date in the coming weeks. During this time, you should write down your reasons for quitting smoking.
2. You should also review behaviors that encourage you to smoke, such as drinking alcohol or being around other smokers, and try to avoid them.
3. Contact your doctor about your plan to stop smoking. Ask him or her if nicotine replacement products or other medications might be helpful for you. Nicotine replacement therapy comes in transdermal patches, nicotine gum, lozenges, vapor inhalers, and nasal sprays.
4. Other non-nicotine medications may help.
5. Get rid of ash trays, cigarettes, and other objects that are associated with smoking.
6. Make your home into a smoke-free zone. If other family members smoke, ask them not to smoke in front of you.
7. Tell your friends and family of your plans to quit, and get their support.
8. Withdrawal from smoking peaks within 1 to 3 weeks after quitting.
9. Be prepared for withdrawal symptoms such as depression, craving for cigarettes, and difficulty concentrating.

Beginning dialysis may be the ideal time for patients to consider quitting smoking. They will be focused on learning healthy lifestyle choices and may be more motivated than when they get more accustomed to dialysis treatments.

## 22. Can my patients exercise?

An exercise program is crucial to feeling stronger on dialysis. Many patients begin an exercise program or physical therapy. They have an increased feeling of well-being, better endurance, and greater muscle strength that allows them to perform many of their most difficult daily tasks more easily. When they stop exercising, many patients feel weaker and more fatigued. An exercise program can make the difference between being an independent patient or one who is reliant on others. The psychological benefits of exercise are well known to athletes and healthcare professionals. Exercise stimulates endorphin release in the brain. Endorphins are chemical messengers released during pleasurable activities. In one study, patients who exercised had similar improvements in their depression as patients on antidepressants.

Remind patients that exercising is also easier to do with others. If they have a regular appointment with a friend to walk, jog, bike ride, or go to the gym, they are more likely to exercise. Patients should always remember: The best results in losing weight are because of changes in a person's lifestyle that result in a gradual loss of weight.

## 23. Can dialysis patients participate in sports?

Dialysis patients can participate in a wide range of sports and activities. Studies have shown that regular exercise and physical activity increase patients' sense of well-being, decrease fatigue, improve depression, lower blood pressure, and improve blood sugar control in diabetic patients. It is important for patients to discuss any sports plans with their healthcare team before starting an exercise program.

Hemodialysis patients with catheters cannot get their catheters wet due to the risk of infection. These patients should not participate in swimming and water sports. Hemodialysis patients with fistulas and grafts can swim and get their access wet. Peritoneal dialysis patients can swim in the ocean or a private pool. Public pools may have a higher bacteria count in the water and are not advised. After swimming, patients on peritoneal dialysis should perform their exit-site care. The peritoneal dialysis catheter should be securely taped to the abdomen during swimming or exercise to prevent irritation of the access site.

Patients on peritoneal dialysis should exercise with their abdomen drained of peritoneal dialysis fluid. They should not perform abdominal exercises such as crunches, sit-ups, or leg lifts. They should perform exit-site care after their workout because sweat will help bacteria grow on the skin.

Hemodialysis patients with fistulas and grafts are discouraged from lifting weights greater than 20 pounds with their access arm. Lifting heavy weights in general is not the best exercise for dialysis patients because

LIFESTYLE CHANGES

lifting has been associated with increasing blood pressure during the lift. Exercises that increase cardiac function such as walking, biking, tennis, and jogging are preferred to exercises that require lifting weights. Calisthenics are good exercises for dialysis patients. After beginning dialysis, many patients feel fatigued and weak and may want to decrease the intensity and level of their normal activities. For example, walking a quarter of a mile daily instead of a mile, or playing tennis for only 10 to 20 minutes rather than an hour. It is not important how much exercise each patient can do; the important point is to start.

It takes effort to overcome the negative feelings of, "I cannot do this anymore," "I'm too old," "People will laugh at me," and so forth. Encourage your patients to overcome these negative feelings. Pass along the following suggestions:

- Be prepared for failure.
- If you become dizzy or feel faint, stop, lie down, and rest.
- Regroup. Ask your doctor about your symptoms. He or she may recommend changes in your exercise regimen that will help you be active without these symptoms.
- Avoid exercising in very high temperatures that can lead to dehydration and heat stroke.
- Use common sense in your exercise.
- If you bike, rollerblade, or ski, wear a helmet (this is good advice for everyone, on dialysis or not).
- Do not swim alone. Use the buddy system.
- Try a new sport.

- Use beginning dialysis as an excuse to treat yourself to something you always wanted to try but never had the time to try. You need to be active to do well on dialysis.

- Take lessons to make sure you are using your sports equipment safely and are getting the most out of your exercise time.

- Have fun!

# 24. My patients wonder how they can keep their bones strong—what should I tell them?

Keeping bones strong requires good nutrition and regular exercise. In addition, kidney dialysis patients are especially susceptible to a type of weakening of the bones known as **metabolic bone disease** or renal osteodystrophy. Bone problems begin early on in kidney failure, before patients begin dialysis treatments. Bones are a large reservoir of both calcium and phosphorus. As kidney function declines, phosphorus levels rise in the blood. The elevated phosphorus combines with the calcium in the blood and pushes the calcium back into the bones. As a result, the level of calcium in the blood decreases.

Another reason for low calcium is decreased vitamin D activity in kidney patients. The kidney is necessary for the production of active vitamin D. Vitamin D helps absorb calcium from the gastrointestinal tract. When the calcium level falls, cells in the parathyroid glands, called chief cells, sense the low calcium level in the blood. PTH travels through the body in the blood, signaling the gastrointestinal tract to absorb more calcium and telling the kidneys to stop eliminating calcium from the body. PTH also tells **osteoclasts** to start releasing calcium

**Metabolic bone disease**

Bone disease in renal patients is caused by reabsorption of bone by cells called osteoclasts whose growth is stimulated by PTH. Patients with bone disease can have bone pain and are more prone to fractures. Also called *renal osteodystrophy*.

**Osteoclasts**

Cells (macrophages) in the bone that remove or resorb bone.

from the bones. If the calcium remains low for a long time, the parathyroid gland grows larger and produces abnormally high levels of PTH. The high PTH level can cause too much calcium to leave the bones, resulting in bones that have less structural material holding us up. If the bones lose calcium, they are more susceptible to fractures. Patients develop changes of the bones of their fingers and collarbones that can be seen on x-rays. The most important evidence that bone problems are occurring is from blood tests. Calcium and phosphorus levels are measured every month in dialysis patients to evaluate bone health. Another blood test, alkaline phosphatase can be elevated if the bone is being reabsorbed too rapidly. PTH is also measured on a regular basis.

The first step in maintaining good bone health is to keep phosphorus levels low. This can be done by decreasing intake of phosphorus-containing foods like milk products and beans. Medications can also be helpful in decreasing the absorption of phosphorus from the gastrointestinal tract. Phosphate binders, which contain calcium acetate, and calcium carbonate binders contain calcium that binds phosphorus. All the phosphorus binders are taken by mouth and trap phosphorus in the gastrointestinal tract. It is important for patients taking these to remember to decrease the phosphorus in their diet. Vitamin D analogs are also frequently used to keep patients' bones strong. They work by increasing the absorption of calcium from the gastrointestinal tract. They also suppress the production and secretion of PTH. Vitamin D is available as capsules taken by mouth and also as an injection that can be given during the hemodialysis treatment.

There is a medication used to treat kidney dialysis patients who are at risk for developing metabolic bone disease that decreases PTH secretion as well as a calcium and phosphorus. This is beneficial because decreasing the amount of calcium circulating in the blood can decrease the calcium that ends up being deposited in blood vessels and other organs of the body.

The concepts of phosphorus and calcium balance may be confusing for patients to understand. Do your best to educate them and continue to emphasize the importance of the renal diet, open communication with healthcare providers, and sticking to regular dialysis treatments.

# PART FIVE

# General Guidelines

Why do some patients do better on dialysis?

What should I tell my patients about the importance of taking their medications as a dialysis patient?

What complications might my patients have?

*More . . .*

## 25. Why do some patients do better on dialysis?

This question of why there is variability in patient outcomes has intrigued the medical community. When two patients who are the same age and have the same diagnosis, blood pressure, cholesterol level, financial resources, and support at home begin dialysis, why does one patient have a better outcome? Why do some patients survive longer on dialysis? This phenomenon has also been noticed in cancer patients, patients with infections such as pneumonia, and in many patients with chronic illnesses. The effect of the psyche or the mind on physical illness should never be underestimated. Patients with a positive attitude do better.

When it comes to dialysis, successful patients are those who are able to commit to the dialysis lifestyle, which can be hard to do. Many patients with kidney failure have neglected their medical care in the past, leading to kidney damage. Patients may have not taken their medication regularly, kept their doctors' appointments, or stayed on their diet. Beginning dialysis treatments is a wake-up call for many patients. Ideally, the message they get is that their health and life are important and that they need to do a better job of self-care.

In some ways, committing to the dialysis lifestyle is like being a competitive athlete. Patients often feel a great deal of loss in beginning dialysis. For instance, their time is not their own. They must show up for treatments at the hemodialysis center or be on a schedule for peritoneal dialysis exchanges. An athlete may not want to go to the pool for 2 hours of swimming laps followed by weight training, but he or she does it in pursuit of a goal. Encourage your dialysis patients to see their treatment the same way. Patients cannot eat whatever they want, just as athletes have to pass up desserts or other treats because they know if they gain weight they will run, bike, or swim slower. By taking the perspective of an athlete, dialysis patients are able to give up their immediate gratification for their long-term goals and are committed to being the best they can be. In truth, life on dialysis is full of choices. While we don't always encourage patients to view it this way, the reality is that they can go to their treatment or skip treatment or decrease treatment time. They can choose to stay on a diet, or eat whatever they want. They can make an effort to remember to take their medication or leave it in the pill container. Successful patients make their treatments time after time. They are not looking for ways to cut corners; they want to be the best they can be. This commitment is not easy or fun, but it will pay off in the long run.

Patients on hemodialysis will do better if they have a good dialysis access. The most common complication in hemodialysis is access problems. Patients with catheters have a high rate of infection and scarring of the blood vessels. Arteriovenous grafts have a high rate of clotting or thrombosis and also are prone to infection, although less so than catheters. The Centers for Medicare and Medicaid Services (CMS) has made lowering the rate of catheter use as a chronic dialysis access one of their Final Measure Specifications for the 2016 ESRD Quality Incentive Program for dialysis centers nationwide. A good, mature AV fistula will help clear more toxins from the bloodstream, thus decreasing patient complications, and will minimize dialysis treatment interruptions due to access failure. A good AV fistula could last 20 years or more. The United States Renal Data System (USRDS) found that patients who had seen a nephrologist for predialysis care were more likely to have a mature AV fistula prior to therapy initiation than those who had not seen a nephrologist for the year preceding treatment.

After access problems, the most common complications in dialysis patients relate to their cardiac health. Heart problems are more common in patients with kidney disease. This is because the risk factors for heart disease such as high blood pressure, diabetes, and elevated cholesterol levels are more common in patients with kidney disease. To do well long-term on dialysis, patients should do what they can to minimize these risks. Patients should be counseled and educated on the following points:

- Smoking: Smoking is one of the most important modifiable risk factors for cardiovascular disease (see Question 21). Patients need support for smoking cessation.
- Hypertension: Whether through diet, medication, or a combination of both, patients need to make blood pressure regulation a priority.
- Diabetes: Diabetes should be properly controlled. Patients may be referred to an endocrinologist if blood glucose cannot be consistently maintained at satisfactory levels.
- Hyperlipidemia: As with hypertension, high cholesterol may be controlled through diet, exercise, and medication.
- Cardiovascular disease: If patients have cardiovascular risk factors, they should be considered for a cardiology referral. An ECG report or a stress test may help with early detection of cardiac problems.

Cardiovascular disease is not the only concern for patients on dialysis. The bottom line is that all aspects of patient health will affect dialysis outcomes. The healthier your patient, the better their treatment will be.

## 26. What should I tell my patients about the importance of taking their medications as a dialysis patient?

Many patients who begin dialysis treatments wonder why they still have to take medication. The average dialysis patient takes between 8 and 12 different pills a day, and taking this amount of medication is difficult for many patients. Dialysis is an imperfect treatment in that the hemodialysis machine and peritoneal dialysis only clean the blood and can remove some medications. Other regular kidney functions (stimulating bone marrow for red blood cell production, activating vitamin D, etc.) are replaced by medications. Patients take medication to regulate blood pressure and lower phosphorus levels. Erythropoietin injections may be necessary if they have a low blood count, as well as injections of vitamin D. Diabetics must continue to take diabetes pills and injections of insulin. Many patients with diabetes need less medication because the kidneys metabolize the insulin the body produces. As kidneys become diseased, there is more insulin available to decrease blood sugar levels. Some patients with type II diabetes can come off their diabetes medications. Patients may take medication to prevent heart attacks, such as beta blockers, ACE inhibitors, and nitrates. An elevated blood cholesterol is associated with kidney failure and needs to be treated with medication when diet alone is not successful. Bowel complaints such as heartburn and constipation are common in kidney patients. It is easy to see how the number of medications adds up quickly.

It is important to emphasize to patients that taking their medications will lower the chances of having a complication that will decrease the quality or length of their life. It is important for patients to learn the reason they are taking each of their medications. The more knowledge they have, the better job they will do taking those pills. Dialysis patients often have several doctors and can end up taking two medications for the same problem. Make sure to look for instances of this when you review patient medications.

Patients may be frustrated with how often they have to take pills. They should be encouraged to ask if they can switch to medicines that can be taken once a day rather than two or three times a day. It may be possible for them to simplify by taking higher doses of one or two blood pressure medications and eliminating the third antihypertensive. Some medications are available in an adhesive patch that is applied to the skin. Patches are especially useful for confused patients who may not remember to take their medication regularly and patients who do not like taking their pills.

Remind patients that if they are very overweight, the best way to reduce their number of medications is to lose weight. Many medical problems such as diabetes, hypertension, and elevated cholesterol get better with changes in diet. Increasing the time spent on dialysis may help normalize blood pressure and phosphorus level. Patients on daily dialysis therapies such as peritoneal dialysis, nocturnal dialysis, and daily home dialysis often are able to stop some of their blood pressure medication and phosphate binders.

## 27. What complications might my patients have?

Feeling weak after hemodialysis treatments is a common experience of some patients. Many patients feel well after their treatments; they finish dialysis and rush off to work, leading a busy life. Other patients find that they must rest or lie down after their treatment but feel much stronger the day following their treatment. When poisons and toxins are removed during hemodialysis, they are removed throughout the body, including the brain. However, there is a lag before they are removed from the brain because the brain and the central nervous system are isolated from the circulation by the **meninges**. The meninges decrease the movement of substances in and out of the central nervous system, forming the blood-brain barrier. When poisons and toxins are rapidly removed from the circulation during hemodialysis, the brain still contains a high concentration of poisons and toxins. Water will move through the blood-brain barrier faster than the poisons and toxins can move out, causing swelling of the brain. This brain swelling occurs to a greater or lesser extent in every patient on hemodialysis.

Brain swelling during hemodialysis is known as **disequilibrium syndrome**. Symptoms of disequilibrium syndrome include headache, weakness, fatigue, nausea, and vomiting. In its extreme form, the patient can become unresponsive or even have a seizure. Fortunately, severe manifestations of the syndrome are rare. In its mild form, it can cause patients to feel washed out and weak. Brain swelling has usually gone down by the next day, which is why patients feel better the day after their hemodialysis treatment. Disequilibrium syndrome does not occur in peritoneal dialysis because peritoneal dialysis is a gentle treatment and removes poisons and toxins

**GENERAL GUIDELINES**

**Meninges**

The three membranes that cover the brain and spinal cord. The meninges form a barrier between the central nervous system and the rest of the body, known as the blood-brain barrier.

**Disequilibrium syndrome**

A syndrome in which the rapid removal of poisons and toxins during hemodialysis results in less toxins being present in the blood than in the brain.

more slowly over a longer time. This allows the brain to equilibrate slowly with the rest of the body, preventing brain swelling. Patients on nocturnal hemodialysis also do not suffer from the disequilibrium syndrome.

Other complications include anemia, hypotension, and muscle cramps. **Table 6** lists complications to watch for in your patients.

## 28. When should I tell my patients to notify the hospital?

The most common reason for dialysis patients to be hospitalized is cardiac problems. Many dialysis patients have risk factors for heart disease such as hypertension, diabetes, hyperlipidemia, and smoking, as well as a positive family history. Therefore, complaints of chest pain should always be taken seriously. Inform patients that chest pain of cardiac origin can radiate to the left arm or neck and can be associated with shortness of breath,

**Table 6**  Potential Complications from Dialysis

| |
| --- |
| Hypotension |
| Hypertension |
| Muscle cramps |
| Anemia |
| Hypokalemia |
| Itching |
| Nausea / vomiting |
| Headache |
| Disequilibrium syndrome |
| Access site complications / infection |

diaphoresis, nausea, and vomiting. This way, if the symptoms occur while the patient is at home, they will know to go to the hospital. A prior history of myocardial infarction, angina, heart failure or arrhythmia should be noted in the patient's record. If chest pain occurs while the patient is having dialysis, vital signs and an electrocardiogram should be taken. The physician covering the dialysis unit must be notified. The immediate treatment for chest pain in the dialysis unit is sublingual nitroglycerin. An 81 mg aspirin tablet should be chewed before being swallowed to improve its absorption to help prevent coronary thrombosis. Oxygen by nasal cannula or mask is given. Rapid transfer to the emergency department will allow further evaluation and treatment in a monitored setting.

The second most common reason for referral to a hospital is infection. Access infection is the most common cause of infection and bacteremia. *Staphylococcus aureus* and *S. epidermidis* ("staph infections") are common pathogens. Some patients can experience infection without a fever. Shaking chills, listlessness or weakness should prompt the dialysis nurse to evaluate the patient's AV fistula, graft, or catheter for sites of infection. Other common sites of infection include pulmonary infections, cellulitis, and foot infections, especially in diabetic patients. After consultation with the dialysis physician, two sets of blood cultures should be obtained before antibiotics are administered. The decision to hospitalize the patient depends on the seriousness of their infection, comorbid medical conditions, and the patient's age.

Arteriovenous access failure used to be the most common reason for admission to hospital. Thrombosis, infiltration, and infection of AV fistulas and grafts remain common problems. Patients with dialysis catheters can have poor or no catheter flow and infection. Dialysis units have developed strategies for treating these problems as an outpatient. Many dialysis units are affiliated with access centers that can treat access problems on a timely basis and even return the patient to the dialysis unit for treatment on the same day. In some units, surgeons, interventional radiologists, and interventional nephrologists can rapidly diagnose and treat access problems as hospital outpatients. Admission to hospital for access problems has become less frequent but still may be necessary. If the patient is fluid overloaded, hyperkalemic, or hemodynamically unstable, admitting the patient may be the safest course of action.

Other reasons for admission to the hospital include gastrointestinal bleeding, abdominal pain, orthopedic injuries and trauma, failure to attend dialysis sessions, and psychiatric disorders. Many dialysis units are unable to give blood transfusions for severe anemia to patients and must refer them to hospital units.

## 29. What role can family members play?

Family members play a key role in supporting patients on dialysis. Hopefully, family members have been involved in the patient's pre-dialysis medical care. It is important for patients to bring a family member, significant other, or friend to their dialysis team meetings to act as a patient advocate. They will receive information from the physician, nurse, dietitian and social worker. The patient advocate can ask questions, clarify information, take notes, and remind the patient what was said.

Family members can help with transportation to and from dialysis, especially at the beginning of dialysis treatments. They can pick up medication from the pharmacy, accompany the patient to medical office visits, and help with shopping, housekeeping and laundry. Family members can participate in a patient exercise program such as walking on non-dialysis days.

One of the most difficult aspects of adapting to hemodialysis is understanding and implementing the renal diet. Members of the family that prepare food should be invited to participate with the patient in meetings with the dietitian. Family members can learn which foods are healthy, which foods to avoid, and integrate these recommendations into the patient's meals. Fluid intake, potassium, sodium, and phosphorus intake need reinforcing at home. Family members often bring important information to the dialysis team concerning what is actually happening at home and can be part of the solution in improving and understanding the patient's diet.

Patients on home hemodialysis and peritoneal dialysis often need significant family assistance with their treatments. Young children and the elderly may depend entirely on family members to perform their home dialysis treatments. Most home dialysis patients receive some help. Setting up the hemodialysis machine, inserting needles, monitoring machine alarms, and cleaning up are some of the tasks in which family members participate. Often, just being available to keep the patient company during their treatments can make home dialysis a successful option for patients.

Family members should participate in the discussion regarding renal transplantation as an option for treatment of kidney failure. They can attend meetings where the risks of surgery and immunosuppression are discussed. This is especially important in the treatment of younger patients on dialysis who may require many decades of renal replacement therapy. As outcomes improve and patients live longer, it is unlikely that patients will be treated with only one modality. It is common for patients to begin on hemodialysis or peritoneal dialysis as their initial therapy. After a period of time they may receive a cadaveric or living-related kidney transplant. Years later, due to chronic kidney rejection, they may be faced with returning to dialysis treatments. Patients on hemodialysis for long periods of time who are experiencing recurrent AV access failure may do better on peritoneal dialysis. The possibility of a successful second, or even third, kidney transplant is a reality for many patients.

Family members will get to know the dialysis nurse, often on a first-name basis. They will experience the hardships of dialysis as well as the treatment successes that are part of patients' daily lives. Integrating family members into the healthcare team is an important step in improving the quality of life of dialysis patients, as well as improving patient outcomes.

## 30. How do I motivate my patients not to give up?

Most patients who are faced with beginning dialysis treatments have negative feelings regarding their treatment and prognosis. They are often depressed and feel that their life is over. Patients are concerned about their ability to work, their relationship with their spouse, and with other important people in their lives. The dialysis nurse plays an important role in helping patients and their families adjust to dialysis. Patients will often share these negative thoughts with the dialysis nurse and other staff members. Listening to the patient in a nonjudgmental way is a good start. Discussions with the patient should acknowledge that beginning dialysis is a life-changing experience and that it is normal to feel depressed.

Patients already on dialysis are important role models for new patients. Ideally, the patient should visit the hemodialysis or peritoneal dialysis unit before beginning treatment. He or she will see patients doing well on dialysis, who are not depressed, and who are more hopeful about dialysis. Asking questions can alleviate many fears and concerns. After meeting with patients, the dialysis nurse can review the dialysis treatment and can explain what will happen on the first day of dialysis. The dialysis unit staff will need to spend a great deal of time and effort with the new patient. The goal is to develop a therapeutic alliance between the patient and the dialysis team. This alliance is a crucial step in helping the patient develop a positive approach to their dialysis treatments. The message to the patient is, "we care about you and we are working hard to make sure that you will do well on dialysis." You also want to convey the message that they *can* do well—that successful treatment is possible to achieve.

Many patients begin dialysis in a debilitated and malnourished state. As uremic toxins are removed and the patient's appetite improves, their physical and psychological condition will improve, and most patients will feel stronger and less depressed. It is important to tell patients that it may take up to six months' time to adjust to dialysis and to get the full benefits of treatment. Praising the patient with phrases such as "I know how difficult this is for you; you are doing a good job" can give patients hope that they can adjust to dialysis and go on with their lives. Hope and confidence are two important components of the patient's struggle with chronic illness. You want to help instill a sense of self-confidence in the patients. Patients who are still depressed after six months of dialysis can often be helped by seeing a psychiatrist or psychologist. Medication for depression and for other psychological disorders is helpful when appropriate.

Physical therapy can increase muscle strength, patient's exercise tolerance, and balance. Patients on dialysis often feel better psychologically after beginning physical therapy; they can become less dependent on others and can participate in more activities. Some patients get support from their church, temple, or mosque. They enjoy being part of a community with a common purpose and feel less isolated when they participate in the many activities that are sponsored by religious organizations. Leaving their job is a cause of loss of self-esteem for many patients. Discussing the possibility of the patient returning to work or working part-time after adjusting to dialysis may be helpful. Some patients enjoy having a part-time volunteer position, which is less stressful but allows them to feel useful and forward-thinking.

End-stage renal failure is fundamentally different from many other illnesses. Patients with heart disease go through a period of diagnosis followed by treatment. They may require cardiac surgery which is painful and physically and psychologically difficult. Six months after surgery they are feeling much better and are back to their routine. Dialysis patients are always facing more treatments. Six months, a year, five years later, they are still having hemodialysis treatments three times a week or peritoneal dialysis treatments daily. They remain on multiple medications, and in the case of hemodialysis patients, still require a restrictive diet. Predicting outcomes for patients is difficult. Renal transplantation, when successful, offers an end to dialysis treatments and a better quality of life. Patients who are doing poorly on dialysis should be referred for a renal transplant evaluation. For many patients, the hope of getting off of dialysis can motivate them to continue their dialysis treatments, improve their compliance with diet and medication, lose weight, and improve their medical condition while still on dialysis. The dialysis nurse plays an important role in educating patients about the value of renal transplantation. Even more important, though, is the nurse's role in thoughtfully assessing the patient's state of well-being and trying to understand each patient's individual needs for information, support, and reassurance.

Most long-time survivors on dialysis have certain characteristics. First, they accept the need for their treatments both intellectually and emotionally. They participate in the planning of their treatments and understand the factors that result in a successful treatment. These long-term survivors take the necessary steps to ensure that they receive a better treatment and resist the temptation to cut their treatment. Second, these patients engage in a healthy lifestyle. They keep their diet, avoid excess alcohol, abstain from smoking, and try to exercise. Third, they are very careful to take their medications correctly. If they have hypertension or diabetes, they control these conditions with diet and medication. By accomplishing these things, they avoid many of the complications of renal failure that interfere with quality of life and longevity. Achieving these goals takes time, and most people are not perfect. By helping patients develop good habits when beginning dialysis, a routine can be established that can promote health and effective treatment for renal failure.

## 31. How can I determine if my patient is compliant in taking their medication?

The average dialysis patient is taking 8 to 12 different daily medications, some of which are taken several times a day. Patients may take medication for diabetes, hypertension, phosphate binders, cardiac medication, and for many other conditions. Multiple physicians will prescribe medications without discussing them with the dialysis team. Keeping track of medications and monitoring patients for side effects are important tasks of the dialysis nurse.

When patients are admitted to the dialysis unit, a list of medications is entered into the medical record. The name of the medication and the reason it is prescribed is reviewed with the patient. The dialysis physician will review the medications and eliminate duplicates, such as two calcium channel blockers prescribed by two different physicians. Many factors contribute to noncompliance with medications. Patients may be on medications they cannot afford or are not covered by their insurance. Often these can be changed to lower-cost medications or medications on the formulary of their insurance. An attempt should be made to decrease the number of medications by eliminating medications that are no longer needed. Prescribing medications that can be taken once daily rather than medications that need to be taken multiple times a day has been shown to improve compliance.

Patients may have cognitive problems due to dementia or stoke. Involving family members or home health aides will help. Patients with persistently high blood pressure or elevated blood sugars need extra attention to make sure they have their medication and are taking it as prescribed. Phosphate binders are frequently not taken by patients due to stomach upset or difficulty swallowing multiple pills. An elevated serum phosphorus level on monthly laboratory testing will alert the dialysis team to this problem. The physician can try changing to another phosphate binder that will be better tolerated. Most medications cannot be routinely measured (although some have blood tests commonly available), and it is up to the dialysis nurse and physician to suspect noncompliance. Asking the patient to bring in the medication rather than the list will uncover patients who need new prescriptions.

Noncompliant patients often state that they are taking their medications to please the dialysis nurse and physician. They give us the answer we want to hear. Many are ashamed of their failure to comply with prescribed medication. The noncompliant patient often provokes feeling of frustration and anger in members of the dialysis team. It is important to realize that many of these patients have been noncompliant for years before they began dialysis. They don't magically become model patients when they begin dialysis. Patterns of behavior are often difficult to change.

The best approach to the noncompliant patient is to find a few extra minutes and pull up a chair next to the patient. In a quiet, private voice, the dialysis nurse should say, "I see that your blood pressure is high today. Is it possible that you may have forgotten to take your medication?" This nonjudgmental approach brings the issue to the patient and expresses concern for their well-being. It is beneficial to try and find a motivation that is important to the patient. For example: "I am concerned about your elevated blood pressure. I am worried that the transplant team will not put you on the active transplant list. Can you bring in your medication at your next treatment so we can work on it together?" Noncompliant patients remain a major concern for the dialysis nurse. It is important for the dialysis team to keep working on the problem in a firm, professional, and caring manner.

The list below shows further examples of questions you can ask patients that uncover the underlying issues surrounding noncompliance with medication.

### Expense

Question to ask: "Were you able to receive coverage for your meds?" This allows you and the patient to engage in a discussion about the cost of taking so many medications. You may be able to provide information about resources that help patients with healthcare costs.

### Schedule

Question to ask: "Did the pharmacy have the medications ready when you picked them up?" Perhaps the patient has difficulty getting to the pharmacy and needs help with transportation. Or maybe there is a home delivery service that would work better for their schedule.

### Frustration/Forgetfulness

Question to ask: "How do you manage to remember when to take each of these pills?" There may be options for a patient to switch to an extended-release version of one or more of the medications they are on, thus reducing the number of pills to take (see Question 26).

### Medication reactions

Question to ask: "How do you feel after you take X medication?" There may be reactions to a certain medication that the patient has not mentioned to any healthcare providers. This is good way to elicit how they physically feel and discuss what symptoms might be concerning and important to report.

The goal is to engage each patient in a discussion about medication in order to reinforce the importance of medication compliance. Helping the patient solve any problems that have become barriers to their compliance is obviously better than potentially causing shame or embarrassment if they have not been taking their meds. There is nothing more rewarding than observing a formerly noncompliant patient's condition improve.

## 32. What educational information should I give my patients?

The National Kidney Foundation (NKF) is a voluntary health organization whose mission is to prevent kidney disease and diseases of the urinary tract. Over the years, the National Kidney Foundation has become a powerful force that has successfully advocated for the kidney patient. Their goals are to support research, train healthcare professionals, expand patient services and community resources, educate the public, and shape healthcare policy.

The American Association of Kidney Patients (AAKP) is a patient-run organization of kidney patients and health professionals. AAKP was founded by six dialysis patients after they met at a Brooklyn Hospital in 1969. From this humble start, this organization has grown into a nationwide group of patients, families, and healthcare professionals that provides support to millions of people with kidney disease. AAKP and its many activities are patient run. AAKP publishes a magazine called *Renalife* that advertises itself as "The Voice of All Kidney Patients." *Renalife* contains articles about dialysis, transplant, diet, travel, and political information written by patients and health professionals.

See the Appendix for more resources available for both patients and nurses.

## *Questions Patients Should Ask Their Healthcare Team*

- Who will help me if I have billing or insurance problems?
- What do I need to do before I start dialysis?
- What do I need to bring to dialysis?
- Is dialysis painful?
- How will I remember everything I'm told at each appointment?
- What should I do if I have questions after hours?
- What should I do if a new symptom arises?
- Will I receive a treatment plan?
- What tests will I need to have done?
- Can I take complementary or alternative medicines?
- What do I do if I need a note for school or work?
- How do I know if I am eligible for Medicaid or Medicare?

## *Questions to Ask Your Dialysis Patients*

- How is traveling to and from dialysis working out for you?

- Do you need refills on any of your medications?

- Does the pharmacy always have your prescriptions ready when you go to pick them up?

- Have you noticed any concerning reactions to your medications?

- Are you taking any complementary or alternative medicines?

- How have you managed making changes to your diet?

- Are you getting enough support from family and friends?

- Do you have any questions for me?

## *How Patients and Nurses Can Stay Current*

### Nursing Resources

**The Institute for Johns Hopkins Nursing**
*http://www.hopkinsmedicine.org/institute_nursing/*
This resource offers consultation and continuing education for nurses.

**American Nephrology Nurses Association (ANNA)**
*http://www.annanurse.org/*
According to its website, ANNA is a community of professional nephrology nurses dedicated to providing the highest quality care. This community offers networking, continuing education, and advocacy opportunities.

### American Society of Nephrology (ASN)

*https://www.asn-online.org/*

The ASN's mission statement includes a commitment to fighting against kidney disease by educating health professionals, sharing new knowledge, advancing research, and advocating the highest quality care for patients.

### American Society of Pediatric Nephrology (ASPN)

*http://www.aspneph.com/*

The ASPN is an organization of healthcare professionals committed to advocating for optimal care for children living with renal disease.

### Nephrology Nursing Certification Commission (NNCC)

*http://www.nncc-exam.org/about/index.html*

The mission of the NNCC is to establish credentialing mechanisms for improving the quality of care provided to nephrology patients.

## Patient Resources

### American Association of Kidney Patients (AAKP)

*https://www.aakp.org/*

As mentioned above, the AAKP is a national community of patients committed to patient-centered education on kidney disease.

### Centers for Medicare and Medicaid Services (CMS)

*http://www.cms.gov/Center/Special-Topic/End-Stage-Renal*
*-Disease-ESRD-Center.html*

The Centers for Medicare and Medicaid Services is a rich resource that provides patients with information on financial assistance (Medicare and Medicaid) as well as a comparison of dialysis facilities and access to a national network of programs that focus on end stage renal disease.

### Mid-Atlantic Renal Coalition (MARC)

*http://www.esrdnet5.org/Home.aspx*

The MARC is one of the Medicare-funded programs committed to patient education and support related to end stage renal disease.

## Needy Meds, Inc.

*http://www.needymeds.org/index.htm*

Needy Meds is an organization that offers information about programs sponsored by pharmaceutical companies designed to help people who cannot afford to purchase their medication.

## National Kidney Foundation (NKF)

*http://www.kidney.org/index.cfm*

A resource for both patients and healthcare professionals, the NKF seeks to prevent kidney diseases, improve the health and well-being of individuals and families affected by kidney diseases, and increase organ availability for transplantation.

## A

**Albumin:** A water-soluble protein found in the blood normally in very minute amounts. Patients with kidney disease often have albumin in their urine because the albumin leaks through the damaged kidneys into the urine.

**Arteriovenous (AV) fistula:** A method for accessing the bloodstream for dialysis that is created by connecting an artery and vein, usually in the arm, using vascular surgery.

**Atherosclerosis:** A disease characterized by thickening and hardening of the arterial wall due to plaque build-up.

**Ascites:** A build-up of fluid in the peritoneal cavity.

## D

**Dextrose:** A sugar added to peritoneal dialysis solutions to help remove fluid from the body. It is also added to intravenous fluids to provide calories for nutrition.

**Disequilibrium syndrome:** A syndrome in which the rapid removal of poisons and toxins during hemodialysis results in less toxins being present in the blood than in the brain. This difference causes water to leave the blood stream and enter the central nervous system, leading to brain swelling. Symptoms include headache, nausea, vomiting, fatigue, and weakness. Most of these symptoms are gone by the next day. A severe disequilibrium syndrome can cause seizures and coma.

**Dry weight:** The body weight of hemodialysis patients at the end of their hemodialysis treatment. The dry weight is determined by removing fluid until the blood pressure begins to fall.

## E

**Erythropoietin:** A hormone produced by the kidneys that signals the bone marrow to produce more red blood cells. In kidney disease, less erythropoietin is produced and anemia can occur. Erythropoietin can be given by injection to treat anemia.

# H

**High-biologic-value protein:** Proteins found in foods such as eggs, fish, poultry, soy products, and meat. High-biologic-value proteins are easily digestible and are rich in the ten essential amino acids that the body cannot make.

**High-flux dialysis:** A procedure that uses high blood flows and large dialysis membranes to remove poisons and toxins in a shorter dialysis treatment, although short dialysis treatments are no longer recommended. The experience with high flux dialysis helped nephrologists improve dialysis treatments by removing more toxins during regular hemodialysis treatments.

# I

**Ideal body weight:** Ideal body weight is dependent upon age, height, body, build, and sex, and is based on the measured average weight of normal individuals. Ideal body weights can be found in published tables, which are available from a kidney dietitian.

**Insensible fluid loss:** The fluid that leaves the body in the form of water vapor from the lungs, sweat, and fluid in stool. The amount of insensible fluid loss is about 2 to 3 cups every day.

# M

**Meninges:** The three membranes that cover the brain and spinal cord. The meninges form a barrier between the central nervous system and the rest of the body, known as the blood-brain barrier.

**Metabolic bone disease:** Bone disease in renal patients is caused by reabsorption of bone by cells called osteoclasts whose growth is stimulated by PTH. Patients with bone disease can have bone pain and are more prone to fractures. Also called renal osteodystrophy.

**Muscle wasting:** A decrease in the size of the muscles of the body due to poor nutrition or medical illness.

# N

**Nocturnal hemodialysis:** Hemodialysis treatments done at night six or seven days a week at the patient's home while the patient is asleep. Nocturnal hemodialysis treatments utilize a lower blood pump flow and a longer duration of treatment time to achieve gentle treatments with better clearance of toxins and poisons.

## O

**Osteoclasts:** Cells (macrophages) in the bone that remove or resorb bone.

## P

**Parathyroid hormone (PTH):** A hormone made and released by the chief cells of the four parathyroid glands found in the neck to regulate calcium metabolism.

**Peritoneal dialysis access:** A surgical procedure in which a plastic catheter is inserted into the abdomen to allow a patient to undergo peritoneal dialysis.

## R

**Recirculation:** Occurs during hemodialysis treatments when blood that has gone through the dialysis machine does not mix with the patient's blood but returns to the dialysis machine.

**Renal diet:** A diet that avoids foods that contain sodium, potassium, and phosphorus, and limits the amount of fluid intake.

**Renal osteodystrophy:** Demineralization of the bone due to renal insufficiency.

## U

**Uremic neuropathy:** Damage to nerves that is caused by the toxins and poisons that build up in the blood in kidney failure.